SpringerBriefs in Applied Sciences and Technology

Nanoscience and Nanotechnology

Series editor

Hilmi Volkan Demir, Nanyang Technological University, Singapore, Singapore

Nanoscience and nanotechnology offer means to assemble and study superstructures, composed of nanocomponents such as nanocrystals and biomolecules, exhibiting interesting unique properties. Also, nanoscience and nanotechnology enable ways to make and explore design-based artificial structures that do not exist in nature such as metamaterials and metasurfaces. Furthermore, nanoscience and nanotechnology allow us to make and understand tightly confined quasi-zero-dimensional to two-dimensional quantum structures such as nanoplatelets and graphene with unique electronic structures. For example, today by using a biomolecular linker, one can assemble crystalline nanoparticles and nanowires into complex surfaces or composite structures with new electronic and optical properties. The unique properties of these superstructures result from the chemical composition and physical arrangement of such nanocomponents (e.g., semiconductor nanocrystals, metal nanoparticles, and biomolecules). Interactions between these elements (donor and acceptor) may further enhance such properties of the resulting hybrid superstructures. One of the important mechanisms is excitonics (enabled through energy transfer of exciton-exciton coupling) and another one is plasmonics (enabled by plasmon-exciton coupling). Also, in such nanoengineered structures, the light-material interactions at the nanoscale can be modified and enhanced, giving rise to nanophotonic effects.

These emerging topics of energy transfer, plasmonics, metastructuring and the like have now reached a level of wide-scale use and popularity that they are no longer the topics of a specialist, but now span the interests of all "end-users" of the new findings in these topics including those parties in biology, medicine, materials science and engineerings. Many technical books and reports have been published on individual topics in the specialized fields, and the existing literature have been typically written in a specialized manner for those in the field of interest (e.g., for only the physicists, only the chemists, etc.). However, currently there is no brief series available, which covers these topics in a way uniting all fields of interest including physics, chemistry, material science, biology, medicine, engineering, and the others.

The proposed new series in "Nanoscience and Nanotechnology" uniquely supports this cross-sectional platform spanning all of these fields. The proposed briefs series is intended to target a diverse readership and to serve as an important reference for both the specialized and general audience. This is not possible to achieve under the series of an engineering field (for example, electrical engineering) or under the series of a technical field (for example, physics and applied physics), which would have been very intimidating for biologists, medical doctors, materials scientists, etc.

The Briefs in NANOSCIENCE AND NANOTECHNOLOGY thus offers a great potential by itself, which will be interesting both for the specialists and the non-specialists.

More information about this series at http://www.springer.com/series/11713

Hilmi Volkan Demir
Pedro Ludwig Hernández Martínez
Alexander Govorov

Understanding and Modeling Förster-type Resonance Energy Transfer (FRET)

FRET-Applications, Vol. 3

 Springer

Hilmi Volkan Demir
Department of Electrical and Electronics
  Engineering, Department of Physics, and
  UNAM—National Nanotechnology
  Research Centre, Institute of Materials
  Science and Nanotechnology
Bilkent University
Ankara
Turkey

and

School of Electrical and Electronic
  Engineering, School of Physical and
  Mathematical Sciences, LUMINOUS!
  Centre of Excellence for Semiconductor
  Lighting and Displays, TPI—The Institute
  of Photonics
Nanyang Technological University
Singapore
Singapore

Pedro Ludwig Hernández Martínez
School of Physical and Mathematical
  Sciences, LUMINOUS! Centre of
  Excellence for Semiconductor Lighting
  and Displays, TPI—The Institute
  of Photonics
Nanyang Technological University
Singapore
Singapore

Alexander Govorov
Department of Physics and Astronomy
Ohio University
Athens, OH
USA

ISSN 2191-530X   ISSN 2191-5318   (electronic)
SpringerBriefs in Applied Sciences and Technology
ISSN 2196-1670   ISSN 2196-1689   (electronic)
Nanoscience and Nanotechnology
ISBN 978-981-10-1874-9   ISBN 978-981-10-1876-3   (eBook)
DOI 10.1007/978-981-10-1876-3

Library of Congress Control Number: 2016943801

Printed on acid-free paper

This Springer imprint is published by Springer Nature
The registered company is Springer Science+Business Media Singapore Pte Ltd.

Contents

Förster-type Resonance Energy Transfer (FRET): Applications

In this chapter, we present several applications of Förster-type nonradiative energy transfer (FRET) related phenomena. In particular, we review light generation and light harvesting applications as well as bio-applications. This chapter is reprinted (adapted) with permission of Ref. [1]. Copyright 2013 Laser & Photonics Reviews (John Wiley and Sons).

1 Introduction

In the past decade, the rise of quantum-confined nanostructures including quantum wires (Qwires), quantum dots (QDs), and quantum wells (QWs) opened new possibilities for various applications. Among these quantum-confined structures, 3D-confined QDs and 2D-confined Qwires are strong candidates for photonic and lighting applications [2, 3]. Colloidal semiconductor QDs are crystalline nanoparticles typically synthesised via wet chemistry techniques [4], with physical dimensions on the order of several nanometres that are generally smaller than or comparable to the bulk exciton-Bohr radius. Therefore, strong quantum confinement effects arise [5, 6]. Semiconductor Qwires are commonly grown using bottom-up techniques via either vapour phase (chemical or physical vapour deposition) [7, 8] or solution-based syntheses (colloidal or hydrothermal) [9]. These Qwires are promising owing to their versatile electrical and optical properties. Qwires can be quite long, on the order of micrometres, characteristically with a small radii ranging from few nanometres to tens of nanometres; therefore, quantum confinement effects can be observed in the radial direction perpendicular to the axial axis.

© The Author(s) 2017
H.V. Demir et al., *Understanding and Modeling Förster-type Resonance Energy Transfer (FRET)*, Nanoscience and Nanotechnology,
DOI 10.1007/978-981-10-1876-3_1

For light-generation and -harvesting systems, semiconductor QDs and Qwires are good alternatives to replace the existing semiconductor thin film materials. To date, QDs have already been utilised as building blocks for light-generation and -harvesting devices [10–12]. QD-based LEDs represent an important class of LEDs that have surpassing performance in the state-of-the-art devices for white light generation [13]. Likewise, Qwires have begun to emerge as auspicious materials for LEDs, lasers, and solar energy-harvesting systems [14–16]. Moreover, Qwire-based LEDs have been shown to be efficient light sources with tunable polarisation and good outcoupling properties. Thus, they have become favourable nanostructures, not only for lighting and harvesting, but also for nanoscale high-speed telecom-munication and computing applications [17, 18]. The photonics properties of the QD and Qwire structures are excitonic in nature; therefore, understanding and mastering their excitonic processes are of high importance for developing advanced and efficient optoelectronic systems employing these materials.

2   Exictonic Interactions with Quantum Dots

QDs exhibit tunable emission spectra, high photoluminescence (PL) quantum yield (QY), broadband absorption spectra and increased environmental stability. These properties have generated significant attraction for QDs to be used in light gener-ation devices. To date, these QDs have been utilised in light-emitting diodes (LEDs) through two primary excitation schemes: (1) colour-conversion LEDs using QDs as colour converting photoluminescent materials [19], and (2) electrically driven LEDs using QDs for electroluminescence via charge injection [11]. In these devices, QDs can be integrated into different material systems including other QDs, QWs, Qwires, carbon nanotubes (CNTs), and organic semiconductors to utilize excitonic processes. Next, we will review the excitonic processes in the various composites of the QDs.

2.1   Quantum Dot—Quantum Dot

Colloidally synthesised semiconductor QDs typically have a finite size distribution, which inhomogenously broadens the emission and the absorption spectra of the QDs. Consequently, excitonic interactions in the distribution of the same QDs arise. Here, we will refer to these as homo-excitonic interactions. The homo-excitonic interactions are important to understand optical properties of the QDs. On the other hand, hetero-excitonic interactions, which occur between QDs of different types, sizes, and compositions, are crucial towards engineering the excitonic operation in the QD composites.

Before discussing the homo- and hetero-excitonic interactions in the QD assemblies, it is worth looking at the effects of different media (i.e., solution phase or solid-state films) on the optical properties of the QDs. In the solution phase, QDs are more isolated from each other, unless the solution is very dense or the QDs are chemically attracted or attached to each other. Therefore, the excitonic interactions between QDs are generally negligible in solution phase. In contrast, when casted into the solid state in the form of close-packed films, QDs are very close with each other, and they consequently exhibit complex excitonic properties. Specific differences between the solution phase and solid-state films are that the photoluminescence emission is red-shifted and photoluminescence quantum yield is reduced in the solid state as compared with the solution phase. The red shift in emission spectra involves both electromagnetic field effects on the transition dipoles in the solid-state state films owing to the substrate and higher refractive index film itself and increased excitonic interactions among the QDs in the form of exciton migration from wider bandgap (smaller size) QDs to narrower bandgap (larger size) ones. First, the substrate and field lead to a change in the effective dielectric medium around the QDs, which causes changes in the spontaneous decay rate and energy of the transition dipole, which is a well-known phenomenon and is not limited to the QDs. Consequently, the radiative lifetime is shortened and the energy of the transition dipole is decreased, which leads to the red shift [20]. Second, the size distribution causes homo-FRET from smaller to larger QDs in the ensemble such that the exciton population in the QDs that are on the red tail of the emission spectrum increases. The reduction in PL-QY is attributed to the increased nonradiative recombination channels that are presented in the solid-state phase. In the solution phase, the QDs are sufficiently apart from each other and the surface traps are effectively isolated in defected QDs. However, in solid-state films, the stacking of the QDs leads to increased exciton transfer and more excitons are trapped in defected QDs, increasing the overall nonradiative recombination.

The homo-excitonic effects have been shown to be important in the exciton migration in the solid-state phase. For example, in the solid-state films of the highly confined silicon QDs, long-range exciton transport was enabled through FRET [21]. When smaller Si QDs were utilised, a longer transport was observed owing to the higher FRET rates. The small QDs facilitate efficient FRET because of their sizes being smaller than the Förster radius. Excitons hop between different QDs multiple times until they are trapped by a large-size QD surrounded with smaller QDs (i.e., a QD with a wider bandgap) [22]. Similarly, it was reported that in QD ensembles, the lifetimes of smaller QDs are shortened due to the exciton transfer to larger QDs, of which lifetimes are increased due to the exciton feeding effect [22–25]. Recently, CdSe/CdS-based QDs have been investigated in terms of their homo-excitonic interactions as a function of the CdS shell thickness. It was found that the homo-exciton transfer in the solid-state films is effectively suppressed due to very thick CdS shells (up to 16 monolayers) in the so-called giant-QDs [26]. As shown in Fig. 1, the emission decay curves of the QD films exhibit large differences at the high-energy and low-energy tails of the emission spectrum as well as at the emission peak, which indicates the occurrence of a homo-exciton transfer for the

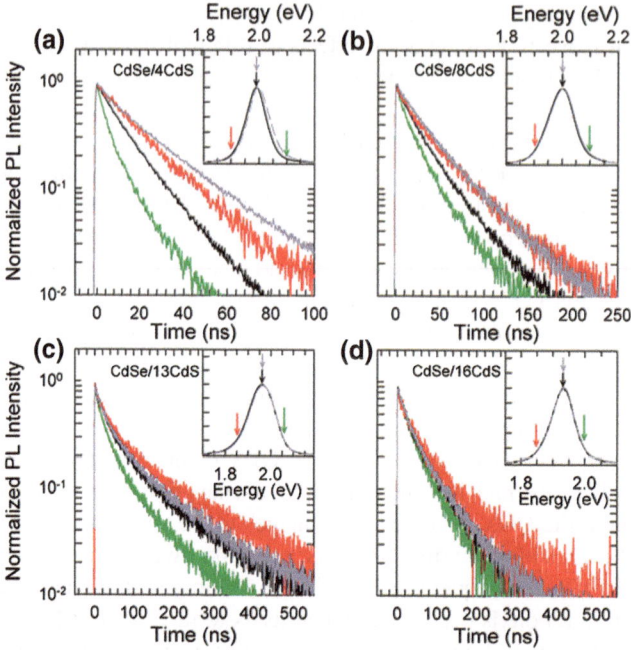

Fig. 1 Time resolved fluorescence decay measurements of the CdSe/CdS QDs depicted with respect to different CdS shell *thicknesses* (i.e., 4, 8, 13 and 16 monolayers). Decay measurements were performed for one QD distribution having only inhomogeneous broadening due to finite size distribution. Measurements were reported at three different spectral positions of the QD emission (i.e., higher- and lower-energy tails and at the peak) in thin film (*green, black* and *red curves*), and also at the peak position in the solution phase of the same QDs (*grey curve*). As the shell *thickness* is increased, FRET process is suggested to be suppressed in the solid-state films of the QDs because the decay curves at different positions of the QD emission spectrum becomes similar. Furthermore, as the shell *thickness* increases, the thin film and solution phase decay curves for the peak position become almost the same, which indicates the isolation of the emitting cores of the QDs owing to the thick shells. Reprinted (adapted) with permission from Ref. [26] (Copyright 2012 American Chemical Society)

thin CdS shells but suppression of the homo-FRET with the giant-shells. As a result, the decay curves measured at different spectral positions of the giant-QDs become indistinguishable.

The hetero-excitonic interactions in the QD-QD structures were investigated for QDs in a wide variety of types, sizes, and compositions. Rogach et al. [27] reviewed examples of the QD-based FRET structures. An exciton in a QD can be transferred to another QD if the donor QD emission spectrally overlaps with the acceptor QD absorption. The transferred exciton rapidly thermalises to the band edge (on the order of ps) in the acceptor such that back-energy transfer is not possible, unless the transfer is coherent due to strong coupling, which is typically not the case for the QD systems. Therefore, excitons have the tendency to migrate towards narrower-bandgap QDs in hetero-structures. The architecture of the hetero-structure plays a crucial role in the

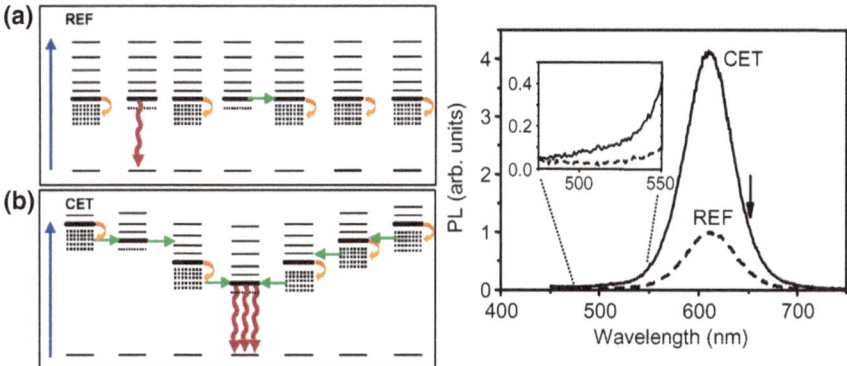

Fig. 2 Two different QD structures are described: **a** non-cascaded reference (REF) structure and **b** cascaded energy transfer (CET) structure. **a** The RET structure consists of layered red QDs. On the *left*, electronic energy levels of the graded QD-employing CET structure and the only *red*-emitting QD-employing REF structure are shown. **b** The CET structure consists of graded layer-by-layer assembled *green/yellow/orange/red/orange/yellow/green* QDs. On the *right*, steady-state PL emission is depicted for both of the structures. The CET structure exhibits substantial enhancement in the PL emission as compared with the REF structure owing to the trapped exciton recycling effect. Reprinted (adapted) with permission from Ref. [28] (Copyright 2004 American Chemical Society)

emerging exciton dynamics. To date, different QD-QD-based structures have been studied in solid-state films using alternative deposition techniques, including layer-by-layer (LbL) [28, 29], Langmuir-Blodgett [30], spin coating [26], drop casting [22], and blending in the polymeric host matrix [31].

Utilisation of layer-by-layer structured QD films with graded bandgap energy was exploited as a means of enhancing the light generation in QDs. This enhancement depends on the recycling of trapped excitons [28, 29]. Figure 2 shows the designed cascaded energy transfer (CET) structure, which is composed of graded bandgap LbL-deposited QDs, and a non-cascaded reference structure (REF) that consists of only red-emitting QDs. In the CET structure, the steady-state PL emission was considerably increased as compared with the REF sample. This enhancement is attributed to the fact that the excitons, which were trapped in the sub-bandgap states of the QDs, can be transferred to narrower-energy-gap QDs. This recycling of trapped excitons leads to a substantial increase in the PL emission of the acceptor QD. This scheme has been applied to colour-conversion-based QD LEDs to enhance the conversion efficiency of the pump photons [31–34].

The rate of exciton transfer in the QD structures has been the subject of several studies [22, 30, 32]. Because of the size distribution of the QD samples, fast FRET rates cannot be ensured in random assemblies of the QDs. However, FRET rates as fast as 50 ps^{-1} with 80 % efficiency were obtained using CdTe QDs with a narrow size distribution in LbL assembled samples [32]. In addition to intrinsic QD properties, organic ligands, which are in charge of passivating the QD surfaces, have also been shown to affect the exciton transfer. Ligands have been shown to change the nature of the transition dipole in the QDs such that higher-order multi-poles should be considered to account for the observed FRET in the QD-QD ensembles [21]. Furthermore, the capability of ligands to passivate the defect and trap sites at the surfaces directly influence the competing exciton transfer rate because exciton decay pathways can be altered via extra nonradiative channels from the surface defects [35].

FRET between QDs has also been investigated from the theoretical point of view [36–39]. Förster resonance energy transfer is considered to account primarily for the observed exciton transfer in QD ensembles due to polydispersity and inho-mogenous broadening effects [36]. However, FRET between single QDs can not be well described with classical FRET. In the case of molecular emitters such as dyes under FRET process, the resonance condition is satisfied by the existence of the spectral overlap between the donor emission and the acceptor absorption. This resonance condition was also discussed for QD-QD assemblies under FRET pro-cess. It was shown that totally resonant or slightly resonant electronic states can perform FRET through direct or phonon-assisted transfer of excitons [37]. Later, two studies questioned the validity of the dipole-dipole coupling approximation for QD structures, and it was shown that the dipole-approximation is valid for donor-acceptor separation distances that are considerably greater than the molecular dimensions [40, 41], therefore, the FRET approach generally provides results that are compatible with the experimental observations.

Recently, Mutlugun et al. [42] have proposed and demonstrated the fabrication of flexible, freestanding films of InP/ZnS quantum dots using fatty acid ligands across very large areas, greater than 50 cm × 50 cm (Fig. 3), which have been developed for remote phosphor applications in solid-state lighting. QDs embedded in a poly (methyl methacrylate) matrix, a myristic acid used as ligand in the syn-thesis of these QDs, imparts a strongly hydrophobic character to the thin film, enables film formation and ease of removal even on large areas, thereby avoiding the need for ligand exchange. When pumped by a blue LED, these Cd-free QD

Fig. 3 Photograph of a 51 cm × 51 cm InP/ZnS QD film under room light along with a ruler (*left*) and the folded film under UV illumination (*right*). Reprinted (adapted) with permission from Ref. [42] (Copyright 2013 American Chemical Society)

films allow for high color rendering, warm white light generation with a color rendering index (CRI) of 89.30 and a correlated color temperature (CCT) of 2298 K. In the composite film, the temperature-dependent emission kinetics and energy transfer dynamics among different sized InP/ZnS QDs were investigated and a model was proposed. High levels of energy transfer efficiency (up to 80 %) and strong donor lifetime modification (from 18 to 4 ns) were achieved. The suppression of the nonradiative channels was observed when the hybrid film was cooled to cryogenic temperatures. The lifetime changes of the donor and acceptor InP/ZnS QDs in the film as a result of the energy transfer were explained well by their theoretical model based on the exciton-exciton interactions among the dots and were in excellent agreement with the experimental results. The understanding of these excitonic interactions is essential to facilitate improvements in the fabrication of photometrically high quality nanophosphors. The ability to make such large-area, flexible, freestanding Cd-free QD films pave the way for environmentally friendly phosphor applications including flexible, surface-emitting light engines.

Also, the authors presented a white LED (WLED), in which both the red and green color components were provided by the green- and red-emitting InP/ZnS QDs forming a bilayer film, as shown in the inset of Fig. 4, designed to result in high photometric quality. Figure 4 shows the resulting emission spectra of the blue LED hybridized with the green-red emitting InP/ZnS quantum dot films and probed using a fiber coupled optical spectrum analyzer. Here, the InGaN/GaN LED was driven at an electrical potential of 4.4 V. The white light generation using the excitation from the blue LED results in a color rendering index (CRI) of 89.30 with a correlated color temperature (CCT) of 2298 K and a luminous efficacy of optical radiation (LER) of 253.98 lm/W_{opt} and hence produces high color rendering, high spectral efficiency, and warm white light. These results demonstrated that these proof-of-concept WLED freestanding films are promising candidates for remote

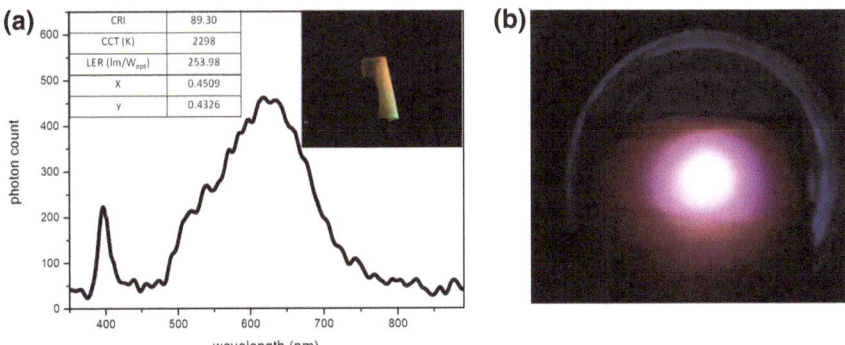

Fig. 4 Electroluminescence spectra of a proof-of-concept white LED using a freestanding InP/ZnS QD film as the remote color-converting nanophosphors together with a *blue* LED chip (The bilayer film consisting of *green* and *red* QDs is shown in the inset). Also an exemplary device under operation is shown on the right. Reprinted (adapted) with permission from Ref. [42] (Copyright 2013 American Chemical Society)

phosphor applications, potentially for high-temperature light engines. [Reprinted (adapted) with permission from Ref. [42]. (Copyright 2013 American Chemical Society).].

2.2 Quantum Dot—Quantum Well

Epitaxially grown QWs are important for optoelectronics, and they have already become the building blocks for various optoelectronic devices such as LEDs, lasers, photodetectors, light modulators, and photovoltaic devices [43]. The current state-of-the-art inorganic LEDs are based on epitaxially grown QWs. These LEDs can be made very efficient, yet it is not easy to tune the emission colour for the generation of white light. The common route to overcome this problem is the utilization of the colour-conversion technique, which relies on a pump LED and colour-converting phosphors. Multiple phosphors (green, yellow, and red) are utilised on top of blue-emitting QW-LEDs to realise the colour conversion. However, these phosphors are limited by their optical properties, such as their broad emission spectra that extend into the far red region in the case of red phosphors, which is spectrally out of the sensitivity range of the human-eye. By contrast, semiconductor QDs exhibit superior optical properties, including a very narrow full-width-half-maximum (FWHM) and tunable emission spectrum in the visible [13]. Therefore, various QD-QW systems have been proposed as efficient colour-conversion materials [44–50] and have recently been reviewed [13].

These QD-integrated colour-conversion LEDs only utilise the radiative energy transfer from the QWs to the QDs. Although high-quality white light generation has been shown to be feasible, radiative energy transfer-based QW–QD colour-conversion systems have some limitations. First, there is a loss mechanism of the pump photons due to the light outcoupling from the high refractive index pump LED into the QD-deposited colour conversion layer. Generally, QDs are encapsulated in a glass-like silicone resin that has a low refractive index. The other limitation is that the nonradiative recombination channels in the pump LED restrict the efficiency of the pump photon usage. To overcome these problems, Achermann et al. [19] experimentally demonstrated an alternative approach where QDs are pumped by QW excitons through FRET in the QW-QD architectures. This type of exciton-pumping was first proposed by Basko et al. [51] for QW-organic emitter system. The proposed exciton pumping of QDs involves the transfer of the excitation energy from the QWs to the QDs where they are in close proximity to each other. To achieve the QDs FRET pumping, an InGaN/GaN-based multi-QW system is used as a working pump LED platform. A GaN capping layer, which was used to passivate the QWs and provide electrical contacts, was thinned to a few nanometres to have an average donor (QW)—acceptor (QD) separation on the order of the Förster radius. With this excitonic pumping QDs, the light outcoupling problem is surmounted because the pump photons are not needed to be emitted into the far field, but are transferred in the near-field via dipole-dipole coupling. Additionally,

FRET creates a competing channel against traps and defects in the QWs such that some of the excitation energy, which was otherwise wasted, could be recycled by transferring it to the QDs (acceptors).

Using this FRET pumping scheme, it was shown that the colour-conversion efficiency can be boosted even by utilising a single monolayer of CdSe QDs on top of InGaN/GaN QWs capped with 3 nm of GaN. The colour-conversion efficiency for this monolayer QD conversion layer was reported to be as high as 13 % [52]. Later, several groups demonstrated that FRET facilitated pumping is not limited to only QD acceptors but organic emitters such as conjugated polymers can also be employed as efficient acceptors [53–58]. A similar scheme was even applied to light-harvesting systems by transferring excitons from QDs to QWs [59–61]. Nevertheless, initial demonstrations of the exciton pumped QD-QW-based colour-conversion LED structures were limited in terms of the FRET rates and efficiencies because of the limited interaction volume between the QDs and the QWs. Although the GaN capping layer could be thinned to make the QWs and QDs closer, the resulting FRET was still restricted because only the top QW and the bottom QD layer could effectively interact. For the other QD and QW layers, FRET was not expected to be efficient due to separation distances greater than 10 nm.

Several groups proposed and demonstrated nanostructured pump LED architectures to promote the FRET between QWs and QDs as opposed to the FRET in the geometrically limited planar architectures [62–64]. These nanostructured pump LED architectures generally employ top-down fabricated nanopillars or nanoholes of the InGaN/GaN multi-QWs. Nizamoglu et al. reported a nanopillar architecture of InGaN/GaN QWs, which are integrated with CdSe/ZnS QDs, resulting in FRET efficiencies up to 83 % for red, 80 % for orange, and 79 % for yellow-emitting QD acceptors [63, 65]. Figure 5 presents a schematic of the nanostructured QW architecture with integrated QDs. A scanning electron microscopy image of the top-down fabricated InGaN/GaN nanopillars, which enable a large interaction volume between the donor and the acceptor species, is also shown in this figure. Furthermore, all the multi-QWs in the pump LED can now contribute to the QDs pumping because the QDs completely surround the nanopillars. In Fig. 5 (bottom), time-resolved and steady-state PL measurements of the QW-QD structure are presented. The exciton decay of the QWs becomes faster upon incorporation with the QDs, which indicates that an efficient FRET channel is created. From the steady-state PL spectrum of the hybrid QD-QW structure, almost totally quenched emission of the QWs can be observed upon introduction of a thin QD layer (several monolayers) on the nanopillar structure.

Recently, exciton pumping in the LbL deposited graded bandgap CdTe QDs on planar InGaN/GaN QWs have been investigated and compared with a non-graded QD acceptor layer. The graded bilayer of the CdTe QDs that consisted of green- and red-emitting QDs (QW-green QD-red QD) exhibited enhanced exciton pumping into the top red QDs (FRET efficiency of 83.3 %) as compared with the reference sample of a bilayer of red-emitting QDs exhibiting much lower FRET efficiency of 50.7 % [66]. The underlying reason was explained via theoretical modelling of the exciton population evolution in the near field. The gradient

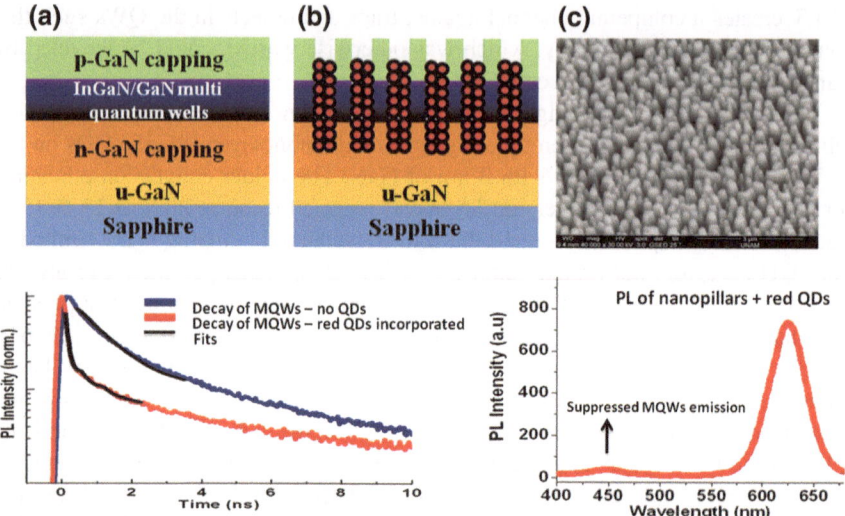

Fig. 5 *Top* schematic illustration of the InGaN/GaN multi-QW architecture and the QD-integrated hybrid. The scanning electron micrograph of the fabricated nanopillar structure is also shown. *Bottom* time-resolved and steady-state PL spectra of the hybrid structure. In the time-resolved PL, exciton decay in the QW was measured before and after the incorporation of the QD. The steady-state PL measurement indicates that the QW emission is almost quenched owing to the efficient FRET. Reprinted (adapted) with permission from Ref. [65] (Copyright 2012 Optical Society of America)

structure enabled faster and unidirectional transfer of the excitons from the QWs into the red-emitting QDs via channelling through the green QDs. In the case of the control sample, the back-and-forth FRET was theoretically shown to slow down the exciton flow from the QW into QDs.

2.3 Quantum Dot—Quantum Wire

QDs integrated into Qwires were demonstrated and investigated for optoelectronics with emphasis on light-harvesting applications owing to the synergistic combination of the strong light absorption properties of the QDs and the superior electrical transport properties of the Qwires. QDs have limited electrical transport properties due to their organic ligands acting as barriers for the carrier transport. Thus, highly conductive and confined Qwires are of great interest as potential hybrid systems, when combined with QDs, for photovoltaics and photodetectors. Kotov et al. investigated semiconductor CdTe Qwires as exciton acceptors, where the colloidal CdTe QDs function as strong light absorber and exciton donor in the specifically functionalised hybrid structure, as shown in the inset of Fig. 6 [67]. As the QDs are integrated into the Qwires, their PL emission spectrum changes, as the excitons are

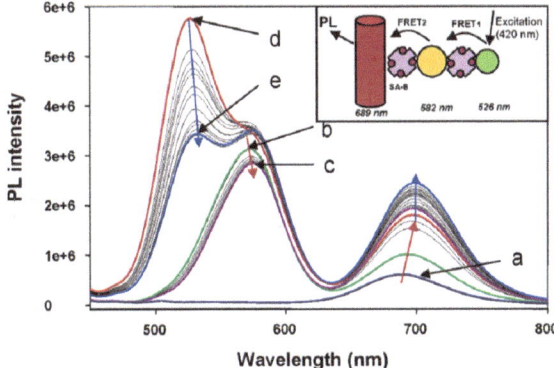

Fig. 6 Exciton energy transfer sensitisation of the CdTe Qwires by CdTe QDs of two different sizes (*orange-* and *green*-emitting) that are specifically attached to the Qwires with an energy gradient structure (Qwire-*orange* QD-*green* QD). Steady-state PL spectra are shown for different cases, which indicates that the emission of the QDs are quenched but the emission of the Qwire is enhanced owing to the exciton funnelling. These systems are promising for excitonically enabled light-harvesting systems. Reprinted (adapted) with permission from Ref. [67] (Copyright 2005 American Chemical Society)

transferred from the QDs to the Qwires. To further enhance the sensitisation of the CdTe Qwires, a cascaded energy system, which consists of green- and orange-emitting CdTe QDs, was utilised. The excitons were efficiently funnelled from the QDs to the Qwires via a two-step FRET process. Later, Madhukar et al. demonstrated a QD-Qwire light-harvesting system and verified that the sensitisation of the Qwires principally occurs via FRET, which was understood through time-resolved photocurrent spectroscopy [59, 68]. Dorn et al. [69] proposed and investigated CdSe/CdS QDs integrated into CdSe Qwires as an efficient exciton-harvesting platform. Furthermore, Hernandez-Martinez and Govorov [70] investigated the FRET dynamics between QD donors and Qwires acceptors with a theoretical model and revealed that quantum confinement of the acceptor Qwire alters the FRET distance (R) dependence to be R^{-5}.

The use of QD-Qwire hybrids towards light generation was also investigated. The transfer of the Qwire excitons into QDs has been realised, especially for ZnO-based Qwires, which can pump the QDs excitonically through FRET [71–73]. ZnO is one of the most suitable materials for this type of excitonic operation owing to its very large exciton binding energy. However, in addition to the proof-of-concept demonstration of the exciton transfer from ZnO Qwires to semi-conductor QDs, the full potential of the exciton pumping via 2D confined structures should be investigated and compared to QW-QD-based schemes.

Another important class of 2D-confined Qwire structures are carbon nanotubes (CNTs). The excitonic nature of CNTs will be discussed in the section on Qwires excitonic interactions. Here, we will describe the QD-CNT-based nanostructures and the underlying excitonic operation. The composite structures of QDs and CNTs have been characterized by several groups, and two reviews highlight the possible schemes of creating hybrid composites of QDs and CNTs [74, 75]. In these composite structures excitonic transfer from the QDs to the CNTs was facilitated, and it was studied through steady-state photoluminescence quenching of the QDs when the QDs are in close proximity to the CNTs [76, 77]. Systematic studies on the separation distance vs. PL quenching of QDs revealed that efficient exciton energy transfer from QDs to CNTs is possible [78]. This exciton transfer increases the photoconductance of the CNTs, which can be beneficial for light-harvesting or - detection systems [79]. Recently, FRET process was accomplished from QDs into several carbon-based nanostructures, including graphene oxide [80], graphite [80], carbon nanofiber [80], and even amorphous carbon thin films [81].

2.4   Quantum Dot—Organics

Colloidal QDs are solution processable materials, which make them compatible with the majority of the organic materials, such as conjugated polymers, dyes, and proteins. These QD-organic hybrid nanocomposites find applications in bio-imaging and sensing, light-emitting devices (LEDs and lasers), and photovoltaics [11, 12, 82, 83]. In addition, such inorganic-organic composites offer rapid and inexpensive processing techniques (roll-to-roll processing), even on flexible substrates. Here, we will focus on the excitonically tailored QD-organic composite material systems. Organic materials have active excitonic properties owing to the strongly bound nature of the excitons, which are called Frenkel excitons. There is a recent comprehensive review paper on the excitonic interactions among organic systems [84].

Integrating QDs into conjugated polymers is a common technique for preparing solid-state QDs films. The excitonic interactions make these nanocomposites particularly important for light generation owing to the possibilities of combining favorable mechanical and electrical properties of the conjugated polymers with excellent optical properties of the QDs. First, Colvin et al. [10] demonstrated a conjugated polymer-QD-based LED which utilised a conjugated polymer as a host charge transporting matrix. Later, exciton transfer from the conjugated polymers to QDs was identified as a possible scheme for light-emitting devices [85]. Spectroscopic evidence of this type of exciton transfer has been reported by several groups. Anni et al. [86] demonstrated that the blue-emitting polyfluorene-type conjugated polymer transfers the optically created excitons into the visible-emitting CdSe/ZnS core/shell QDs via FRET. Similarly, exciton transfer was reported for infrared-emitting PbS QDs integrated with different conjugated polymers [87–89]. Following these initial reports, several studies have focused on developing a deeper understanding of the excitonic processes between conjugated polymers and QDs [90–92].

Stöferle et al. [93] demonstrated that diffusion of the exciton in the conjugated polymer is a vital process for FRET to occur from conjugated polymers to QDs, especially at low QD loading levels in the polymeric films. Lutich et al. [94] revealed the excitonic interactions in an electrostatically bound QD-conjugated polymer hybrid in solution phase; although there is a type II band alignment in the QD-conjugated polymer composite, the dominant excitonic process is found to be FRET rather than charge transfer or Dexter energy transfer process. Figure 7 presents the time-resolved fluorescence decay of the donor polyelectrolyte poly[9,9-bis (3′-((N,N-dimethyl)-N-ethylammonium)-propyl)-2,7-fluorene-alt-1,4-phenylene] dibromide (PDFD) polymer and acceptor CdTe QDs that have negatively charged ligands before and after the integration in the solution phase. The PDFD conjugated polymer has a single exponential lifetime in the absence of the acceptors, but a double exponential fit could only account for the measured decay curve in the presence of the acceptors. The newly appeared decay path has the same lifetime scale as the exciton feeding process in QDs (see Fig. 7 top), which confirms that the excitons are transferred from the PDFD to the QDs. The efficiency of the FRET process was measured to be 70 %. Ultimately, the interaction zone of the long-range FRET and short-range Dexter energy transfer can be seen in Fig. 7 (bottom).

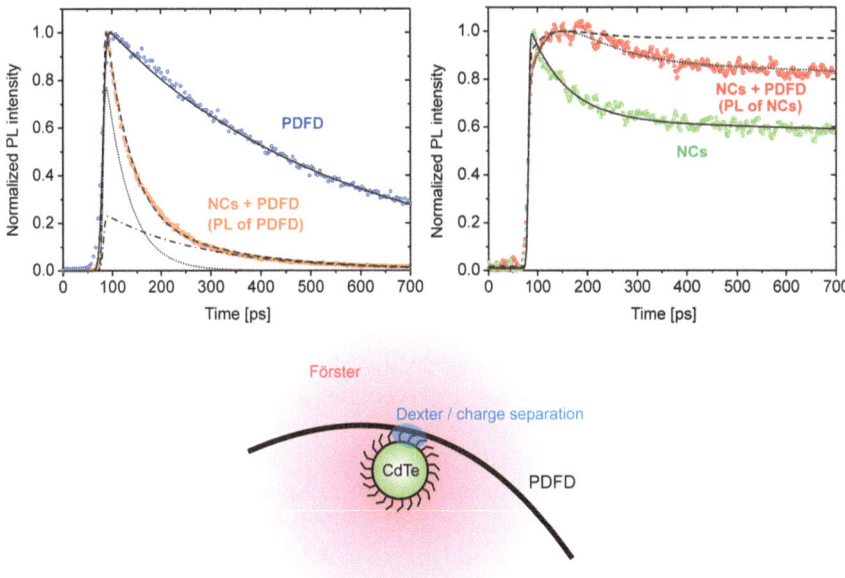

Fig. 7 Time-resolved fluorescence decays for the donor PDFD and acceptor CdTe QDs in the PDFD-CdTe QD hybrid nanocomposite (solution phase) are shown before and after incorporation. The decay of the PDFD becomes significantly faster upon QD integration owing to the efficient FRET. The decay of the QD shows the exciton feeding on the same time scale of the FRET via slowing in the decay curve. Although there is a type-II band alignment in the nanocomposite, the dominant excitonic interaction is FRET with 70 % efficiency. Other excitonic interactions, such as Dexter energy transfer and charge separation, are limited due to their short range operation, as shown in the bottom schematic of the hybrid. Reprinted (adapted) with permission from Ref. [94] (Copyright 2009 American Chemical Society)

The exciton transfer dynamics can be modified by the architecture of inorganic-organic nanostructure. For example, a LBL deposited hybrid assembly of CdTe-QDs and polyelectrolyte conjugated polymer showed suppression of nonradiative channels in the polymer [95]. Furthermore, in the conjugated polymer-QD mixtures, one important effect that should be considered is the phase segregation of the constituent materials. This segregation is observed in the mechanically blended QD-conjugated polymer systems such that the QDs tend to form aggregates in the solid-state films. The phase segregation restricts FRET in the QD-polymer films via suppressing the interaction volume. Therefore, it is crucial to control the nanoscale interactions in these hybrids to achieve the desired excitonic operation [96–99].

Small organic molecules are frequently employed in organic LED (OLED) and organic photovoltaic (OPV) devices as electron-hole transport or emissive layers. Furthermore, these molecules are also employed in QD-based LEDs; therefore, it is important to understand the excitonic interactions between these small organic molecules and QDs to engineer QD-based LEDs [11, 100, 101]. The charge injection from the adjacent organic layers into the QDs is not efficient due to unbalanced injection leading to Auger recombination in the QDs [102]. By contrast, excitonic injection could resolve this charging issue and subsequent Auger recombination problem. Therefore, maximising the excitonic injection from the adjacent small organic molecule layers into QDs is vital.

For example, TPBi (1,3,5-tris(N-phenylbenzimidizol-2-yl)benzene), which is one of the most frequently used electron transport and hole blocking layer, was shown to possess an exciton transfer efficiency up to 50 % into core/multi-shell CdSe/CdS/ZnS QDs [103]. The engineering of the shell composition and thickness to match with the TPBi emission was shown to enhance excitonic interactions. Later, TPD (N'-diphenyl-N, N'-bis(3-methylphenyl) 1, 1'-biphenyl-4, 4' diamine) and TcTa (4,4',4''-Tri(9-carbazoyl)triphenylamine), which are widely used for hole transport purposes, were also shown to offer a large exciton transfer capability when they are adjacent to QDs [104]. Additionally, phosphorescent molecules, where heavy metal atoms create a strong spin-orbit coupling and intersystem crossing, have highly emissive triplet-states. These phosphorescent molecules are also promising candidates for exciton injection to QDs. It was demonstrated that an iridium complex phosphorescent molecule called Ir(ppy)$_3$ (fac-tris(2-phenylpyridine)iridium) can enhance the steady-state PL emission of the CdSe/ZnS core/shell QDs in a bilayer film structure of QDs and Ir(ppy)$_3$ in CBP (4,4'-N,N'-dicarbazolyl-1,1'-biphenyl) [105]. However, the underlying physics of the exciton transfer between the QDs and organic molecules is still unclear, whether the main transfer route is through FRET or Dexter transfer. Nonetheless, this scheme was applied to hybrid QD-LEDs and slightly better enhancements were observed in the external quantum efficiencies (EQEs) of the devices [106–108]. Although there are concomitant enhancements in the performance of these hybrid devices, the efficiencies are still well below the EQEs of those devices made of only phosphorescent materials (>20 % EQE). More suitable architectures, rather than simple bilayers of the QDs and phosphorescent molecules, are desired for efficient excitonic operation.

Similarly, bio-conjugates of the QDs with proteins have been investigated for imaging, labelling, and sensing applications in biology [82]. These QD bio-conjugates are excitonically active such that the QDs can function as both the exciton donor and acceptor [109–114]. These hybrids can be promising for future lighting and light-harvesting systems. For example, chemical and biological systems can produce light upon molecular-level interactions via the so-called chemiluminescence and bioluminescence. These systems can be used as novel light generation structures with the incorporation of QDs owing to their superior colour control and tuning abilities. Through bio- or chemi-luminescence resonance energy transfer (BRET or CRET), the excitation energy created in the bio- or chemi-luminescent system can be transferred to the QDs [115–117]. The initial demonstrations for chemical and biological systems were commonly targeted for applications as external light sources for bio-imaging and sensing applications. Electrically activated chemiluminescence systems that can transfer excitons to QDs were shown to be favourable for sensing applications [118, 119]. In addition to bio-conjugates, QD-dye hybrids also show promise for biological sensing and labelling applications. FRET between luminescent dyes and QDs have been studied in detail to elucidate the effects of concentration, shape and structure of the hybrids [120–124].

3 Excitonic Interactions in Quantum Wires

The excitonic operation is also prevalent in semiconductor Qwires, and many groups have studied the excitonic properties of various Qwire systems in the pursuit of obtaining a better understanding of the photonic properties of the Qwires. Excitons in the Qwires are confined in two dimensions, and optical properties of these excitons are generally less pronounced as compared to those observed in QDs. Because it is not easy to fabricate Qwires with diameters smaller than 10 nm with the available physical and chemical vapour deposition techniques whereas colloidally synthesised QDs can be made quite small—on the order of a few nanometres in diameters. Therefore, QDs exhibit much stronger quantum confinement than the current Qwires do [125–127]. To attain the large binding energies required for creating stable excitons, the physical dimensions of the Qwires should be made smaller than the bulk exciton Bohr radius.

To date, Qwires of a broad range of semiconductor materials have been synthesised and their excitonic features have been confirmed and investigated using optical spectroscopy. In Qwires, which have poor quantum confinement and small bulk exciton binding energy, excitons are not stable and they are easily dissociated into free carriers at room temperature (i.e., $k_B T \sim 25$ meV $> E_B$). Therefore, the Qwires typically need to be cooled down to observe the excitonic features in their optical properties. However, materials with large bulk exciton binding energies such as ZnO, ZnS, and CdS can exhibit room temperature excitonic behaviour even under weak quantum confinement. Furthermore, for Qwires, for example, made of CdSe, InP, and GaAs, exciton binding energies are nearly an order of magnitude

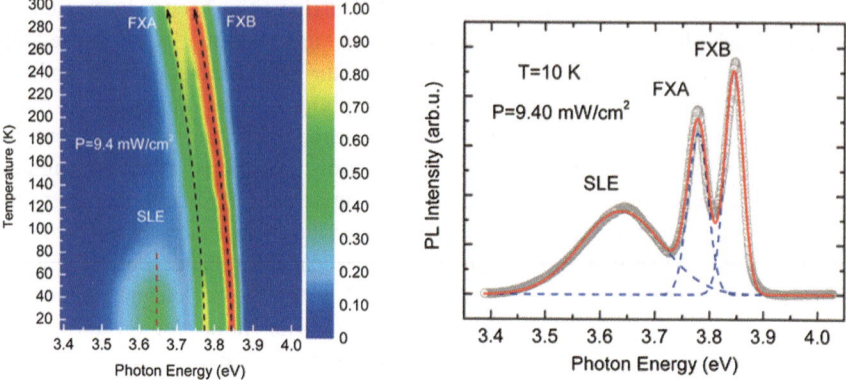

Fig. 8 Excitonic emission mapping of the ZnS Qwires extracted from temperature dependent steady-state PL measurements. These Qwires exhibit efficient excitonic operation even at room temperature owing to the large exciton binding energy of ZnS. On the right, steady-state PL emission at 10 K is shown. FXB and FXA are the free exciton states of A and B separated by 57 meV. SLE is the shallow level emission due to defects. Reprinted (adapted) with permission from Ref. [129] (Copyright 2010 American Chemical Society)

larger than the bulk binding energies owing to the quite small radii (<5 nm) of the Qwires [126, 127].

The crystal quality and defects are important for the optical and excitonic properties of the Qwires. Nonradiative relaxation channels due to the presence of defects should be suppressed. Therefore, it is crucial to have high crystal quality and low defect densities [128]. For example, the temperature dependent optical properties of ZnS Qwires were investigated by Chen et al. Figure 8 presents the excitonic emission spectrum at 10 K with the mapping of excitonic emission peaks as a function of the temperature. Excitonic operation, which is essential to maintain the efficient radiative recombination of the optically pumped ZnS Qwires, was shown to be apparent at room temperature owing to the large exciton binding energy of the ZnS Qwires. These ZnS Qwires are promising new materials for UV-emitting LEDs and lasers [129].

One promising use of the excitonic phenomena in Qwires is the stimulated emission generation and lasing. Since excitons are bound entities and tend to radiatively recombine, they lead to strong light emission properties in materials. In combination of the excitonic nature and the unique structural advantages of the Qwires such as light confinement, Qwire systems were shown to be suitable for lasing applications. Huang et al. [130] first showed the applicability of ZnO Qwires as an active gain medium for optically pumped random nanolasers. Lasing was exhibited at room temperature from the fabricated ZnO Qwires, which show strong exciton and photon confinement properties. In 2005, Agarwal et al. [131] demonstrated strong lasing emission from high-quality single-crystal CdS Qwires. CdSe Qwires can also be utilised as room temperature lasing media at near-IR wavelengths. Although the excitonic operation was strong at room temperature in these Qwires, weak phonon-exciton interactions were demonstrated to assist the room temperature lasing [132]. Xu et al. revealed an amplified spontaneous emission (ASE) build-up in a network of one dimensional CdS-based nanobelts. The defects in the nanobelts revealed that they alter the excitonic processes by creating bounded excitons. Engineering these defects was shown to provide fine control of the ASE threshold due to the interplay between the bound exciton density and Auger recombination kinetics [133]. Qwires, which have a naturally formed flat facet, are also good candidates for optical amplification as semiconductor gain media [15]. In addition to laser diodes, Qwires have also been employed in LEDs, photodetectors and FETs [18].

Structured Qwires have also been introduced for obtaining high-quality and functional 1D systems. In these Qwires, alternating materials (i.e., at least two or more) are grown either on the axial direction (end to end stacking) or on the radial direction (core/shell like stacking) [134]. These hetero-structured Qwires can be made of p-n junctions on single wires. Furthermore, core/shell Qwire hetero-structures have shown advantageous properties for excitonic control because the cores can be highly confined and passivated via shell growth such that 1D excitonic operation can be efficiently preserved. Recently, Qian et al. demonstrated GaN/AlGaN multi-core/shell Qwire hetero-structures, in which the GaN core is surrounded by highly uniform GaN/AlGaN multiple QWs shell (Fig. 9a) [3, 135, 136]. The TEM studies revealed that the growth of multiple QWs based on a GaN core is epitaxial and dislocation-free. The emission and lasing wavelength of the multi-core/shell Qwires, which is determined by the AlGaN component, can be tailored over a wide-range at room temperature (365–494 nm), as shown in Fig. 9e. In addition, the photon confinement and, consequently, the mode volume in the GaN core can be tuned by the number and structure of the QWs shell. This hetero-structure is suitable as a lasing medium owing to the exciton and photon confinement effects.

As another technologically important material platform, zinc oxide is an emerging semiconductor material with a very high bulk exciton binding energy (~ 60 meV). Therefore, the excitonic features can be easily observed in the optical properties, even at room temperature. This property has led to the development of ZnO-based optoelectronic devices over the last decade. ZnO Qwires were employed in LEDs by combining different materials such as p-Si and p-GaN with n-ZnO Qwires [137–142]. Zimmler et al. [143] demonstrated a single ZnO Qwire LED, where EL spectrum was investigated as a function of the temperature (7–200 K), and at low temperatures, the emission spectra of the LED was dominated by strong excitonic emission. Recently, using the piezoelectric characteristics of ZnO, mechanical deformation has been shown to modify the excitonic features in the emission spectra [144, 145].

3.1 Excitonic Interactions in Carbon Nanotubes

Carbon nanotubes (CNTs) make an interesting class of 1D materials that exhibit unique Qwire properties. A recent review on the single-walled CNTs (SWCNTs)

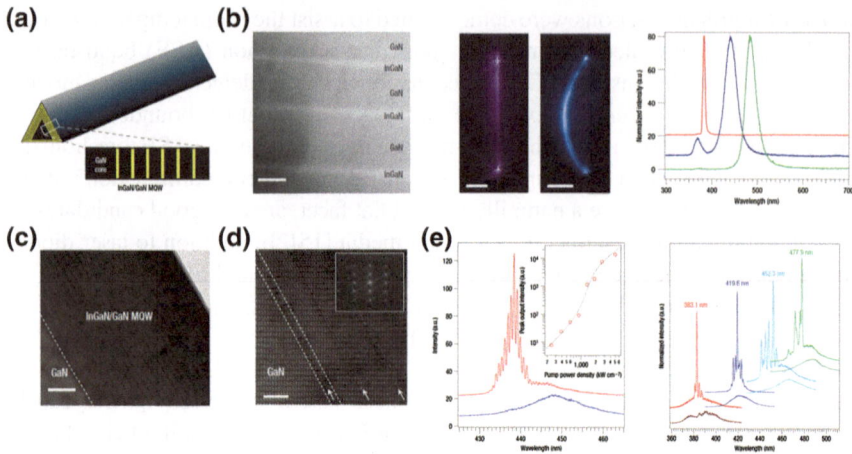

Fig. 9 a Schematic diagram of GaN/AlGaN multi-core/shell Qwires. The core is composed of GaN Qwires. The shell is composed of multiple GaN/AlGaN Qwells. **b** *Dark-field* cross-sectional STEM images of multi-core/shell Qwires with 3 Qwell layers. The scale bars are 20 nm. **c** *Bright-field* TEM image of a multi-core/shell Qwires with 26 Qwell layers. The scale bar is 10 nm. **d** Enlarged TEM image of MQWs in **c**. White arrows highlight InGaN quantum-well positions. The *dashed line* outlines the interface between an InGaN quantum well and adjacent GaN quantum barriers. The scale bar is 2 nm. Inset: Two-dimensional Fourier transforms of the entire image. **e** Emission and lasing properties of as-grown multi-core/shell Qwires. The PL image and PL emission band varies as the component of shell (*left-right*, *upper*). With GaN/AlGaN Qwells shell as gain media and GaN core as cavity, the multi-core/shell structures can be serve a micro-laser (*left*, *bottom*). The lasing wavelength is effectively tuned from 365 to 484 nm (*right*, *bottom*). Reprinted by permission from Macmillan Publishers Ltd: Nature Materials (Ref. [135]), copyright (2008)

summarises the excitonic properties in the CNTs [146]. Experimental and theoretical studies have shown that the optical properties of CNTs are governed by strong excitonic behavior [147–149]. For example, the excitons in semiconducting single-walled CNTs (SWCNTs) were observed to be strongly bound owing to the large exciton binding energies ($\sim$300–600 meV) [150], and these excitonic features are dominant in the CNTs' optical absorption [149]. This active excitonic operation in SWCNTs makes them auspicious materials for light-harvesting applications. Consequently, semiconducting SWCNTs were employed as near-IR light harvesters by hybridising them with C_{60} molecules in a bilayer architecture. Exciton dissociation was demonstrated at the proposed CNT-C_{60} interface, although CNT excitons have quite large binding energies [151]. To assess the device's performance, carefully sorted semiconducting SWCNTs were used as the active absorber layer with a film thickness less than the exciton diffusion length of the CNTs such that excitons can easily become dissociated [151]. In addition, exciton diffusion and mobility of the excitons were measured in the SWCNTs via photoluminescence quenching experiments, and the results revealed that the exciton diffusion lengths are up to 250 nm, which is along the nanotube axis [152].

With the goal of enhanced light-harvesting systems based on CNTs, S. Wang et al. investigated multi-exciton-generation in the SWCNTs via high-energy photon irradiation (i.e., 355 and 400 nm). The absorption of the high-energy photons was shown to create multi-excitons in the semiconducting SWCNTs with a carrier-multiplication threshold that is close to the theoretical limit ($h\upsilon \sim 2E_g$) [153]. The Auger-type exciton-exciton annihilation processes were shown to be highly effective in these 1D confined CNTs owing to the strongly bound nature of the excitons, similar to the case of the QDs [153]. However, Auger annihilation is very fast ($\sim$ tens of ps) and it may limit the effectiveness of light-harvesting systems due to the loss of excitons through thermalisation. On the other hand, SWCNTs were observed to emit light via the exciton radiative recombination channel [154]. However, not all the excitons can decay radiatively because of the existence of the so-called dark excitons [148]. The infrared emission from the SWCNTs was later employed for electroluminescent devices [155–157]. The external quantum efficiencies of these proposed devices were measured to be on the order of 10^{-4}. This poor performance was attributed to the poor PL QY of the CNTs, which is on the order of 0.01 [155]. Recently, the use of asymmetric contact was proposed to enhance the CNT-based LEDs, which were shown to exhibit narrow excitonic emission at 0.9 eV [156, 157].

In addition to these excitonic features in the optical properties of the CNTs, the inter-CNT excitonic interactions (exciton transfer) have also been investigated in literature. Exciton energy transfer in the form of long-range FRET was shown in the bundles of the SWCNTs [158–161]. Figure 10 shows the mechanism for FRET in the SWCNT bundle. A SWCNT with a larger band gap can transfer exciton to another SWCNT that has a smaller band gap. The FRET efficiency is given as a function of the separation distance between the donor-acceptor CNTs (Fig. 10). For distances equal to or less than 3 nm, efficient FRET could be observed. The primary limitation behind the observed small Förster radius (2–3 nm) was attributed to the

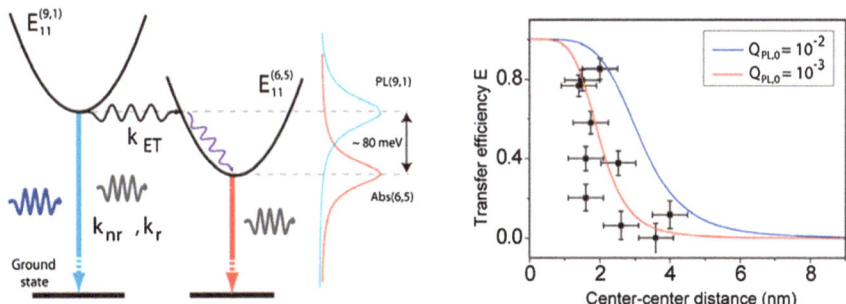

Fig. 10 Exciton transfer via FRET in the SWCNT bundles. Similar to FRET in QD assemblies, FRET can occur in the CNT bundle from a larger-bandgap CNT to a smaller-bandgap CNT. Energy-transfer efficiency is plotted as a function of the separation distance between the two interacting CNT. Reprinted (adapted) with permission from Ref. [159] (Copyright 2008 American Chemical Society)

very low photoluminescence of the CNTs, which makes them inefficient exciton donors. In addition to SWCNTs, exciton transfer was observed in the double-walled CNT (DWCNT) assemblies, where the energy transfer occurs in both intra- and inter-CNTs [162, 163].

4 Excitonic Interactions Beyond the Förster Limit

4.1 Plasmon—Exciton Interactions for Enhanced Excitonic Coupling (Plexcitons)

Plasmonics is an emerging field in nanophotonics and has applications ranging from solar cells to photonic devices. Plasmons consist of the collective oscillations of free electrons in metals. Investigation of the plasmon-exciton couplings (also called plexcitonics) has created a great interest in the scientific community. Several groups have studied fluorescent materials attached to metallic nanoparticles in efforts to elucidate this complex interaction [164–168]. It was shown that nearby plasmonic oscillations modify the radiative recombination rate of the fluorescent material. Moreover, the emission of an emitter can also be enhanced owing to the enhancement of the radiative rate over the nonradiative ones. Furthermore, when the separation is sufficiently small (sub-5 nm), nonradiative energy transfer becomes dominant. This nonradiative energy transfer leads to the strongest energy transfer from the fluorophores to the plasmons; consequently, the fluorophore emission is severely quenched. A considerable number of experimental works have reported that the intensity of emission in this sub-5 nm region monotonically varies with decreasing separation [169–172]. Recently, Peng et al. [173] have shown that in a core-shell/multishell plasmonic nanocavity the spectral overlap between the plasmon and the emission bands also plays an important role in the energy transfer process, which can lead to a minimal emission intensity at ~ 2 nm.

For plasmonically enhanced emission from semiconductor nanocrystals, the exciton-plasmon and photon-plasmon resonance conditions are important, and can be formulated in the following simple way [164]:

$$\omega_{exciton} \approx \omega_{plasmon} \text{ and } \omega_{photon} \approx \omega_{plasmon} \qquad (1)$$

where the frequencies involved are related to excitons in a semiconductor component $(\omega_{exciton})$, plasmons in metallic components $(\omega_{plasmon})$, and photons of incident light (ω_{photon}). Under illumination of a given intensity, an exciton-plasmon nanocrystal complex constructed by using the above conditions (1) can exhibit strongly-enhanced emission. Under the exciton-plasmon resonance $(\omega_{exciton} \approx \omega_{plasmon})$, the enhancement results from an increased probability for an exciton to emit a photon since an exciton is coupled with a plasmon and, in this way, acquires an enhanced optical dipole. Under the second condition $(\omega_{photon} \approx \omega_{plasmon})$, excitons inside a hybrid

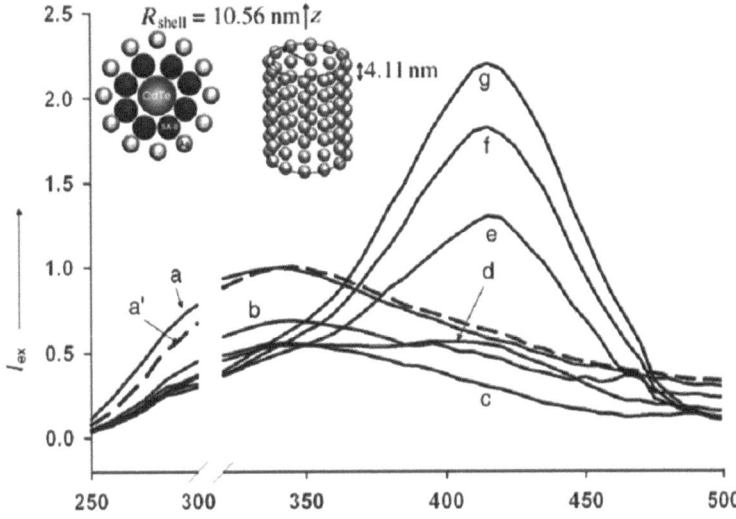

Fig. 11 Schematics of the CdTe-Ag nanowire-nanocrystal structure with enhanced emission properties due to the photon-plasmon resonance. The structure was assembled in a solution using special bio-linkers (SA-B). The plot shows the photoluminescence excitation (PLE) spectra at the peak emission wavelength of the CdTe nanocrystals. In every 10 min, PLE spectrum is measured for $a \rightarrow g$. As the CdTe-Ag hybrid is formed in solution, the significant enhancement in PLE signal is observed at ~ 420 nm. The schematic shows a cross-section with a central CdTe NW and an Ag-nanoparticle shell. Reprinted by permission from Wiley (Ref. [177]), Copyright © 2006 WILEY-VCH Verlag GmbH & Co. KGaA, Weinheim

semiconductor-metal system acquire an increased absorption cross-section that also leads to amplified emission. Interestingly, semiconductor emitters and metal plasmonic amplifiers can be made of various shapes and dimensionalities. This is archievable because of the wide variety of possibilities enabled by the modern nanofabrication and synthesis techniques today. An example is the nanocrystals bio-assembly in a liquid phase [174, 175]. Two other examples consisting of such nanowire-nanocrystal structures (CdTe-Au and CdTe-Ag) with strongly enhanced PL emissions are reported and described in Refs. [176, 177]. Different metals can sustain different surface plasmon resonances, e.g., at ~ 500 and ~ 400 nm for Au and Ag nanocrystals, respectively. These plasmon resonances were employed for a realization of the two resonance conditions given in (1), using CdTe Qwires as emitters. Consequently, the structures designed according to these conditions worked well as plasmonic amplifiers for the exciton emission from the semiconductor Qwires. Figure 11 shows a nanostructure complex used to investigate the photon-plasmon resonance, which involves CdTe nanocrystals and Ag nanoparticles [177]. As the CdTe-Ag hybrid system is formed in solution via bio-linkers, the photoluminescence excitation spectra (PLE) at the CdTe peak emission wavelength exhibits strong enhancement for the spectral region around the plasmon resonance of the Ag nanoparticles as shown in Fig. 11.

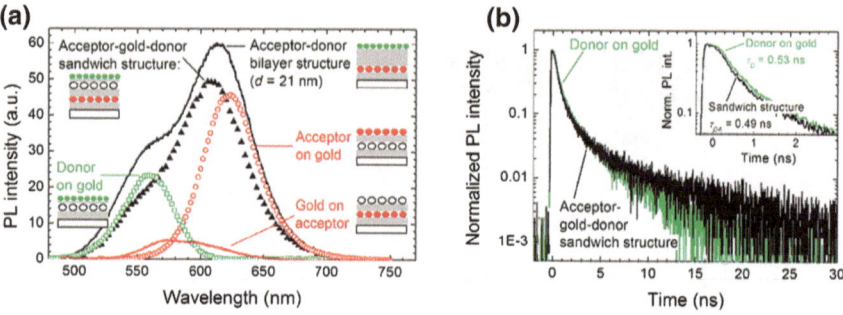

Fig. 12 Enhancement of FRET in the CdTe QD—Au nanoparticle layer-by-layer coated system investigated using **a** steady-state and **b** time-resolved spectroscopy. **a** Using four different negative control samples (donor on gold, acceptor on gold, gold on acceptor and acceptor-donor bilayer structure) and the working sample (acceptor-gold-donor sandwich structure) steady state emission properties of the QDs were compared. **b** In the time-resolved decay curves, modification of the exciton lifetime of the donor QDs at a large separation distance is shown in the presence and absence of the gold nanoparticles. The acceptor-gold-donor structure exhibits faster donor lifetime as compared to the case of donor on gold, which is attributed to the enhanced FRET owing to the gold nanoparticles. Reprinted (adapted) with permission from Ref. [180] (Copyright 2011 American Chemical Society)

FRET between quantum confined structures can also be modified in the presence of plasmonic coupling. The phenomenon of the FRET enhancement via plasmonic coupling was theoretically described for the case of QDs by A. O. Govorov et al. [178]. Later, it was experimentally demonstrated that localised plasmons could enhance the dipole-dipole coupling. In turn, FRET can be enhanced, even at larger separation distances [179, 180]. Figure 12 presents the steady-state and time-resolved signatures of the FRET enhancement via localised plasmon oscillations of Au nanoparticles. This enhancement was corroborated via both spectroscopic measurements and detailed modelling. The Förster radius was observed to almost double owing to the antenna effect of the Au nanoparticles for the case of FRET from green-emitting to red-emitting CdTe QDs in a LbL assembled architecture. Furthermore, using a plasmonic cavity coupled to QDs, polarised emission was detected from isotropic QD emitters, which is very promising for polarised light generation [181–183].

4.2   Effect of Plasmonic Coupling on Förster-type Nonradiative Energy Transfer

In this work, Ozel et al. [184] reported selectively plasmon-mediated nonradiative energy transfer between donor-acceptor (D-A) quantum dot emitters interacting via Förster-type resonance energy transfer under controlled plasmon coupling either to only the donor QDs site (i.e., donor-selective) or to only the acceptor QDs site (i.e.,

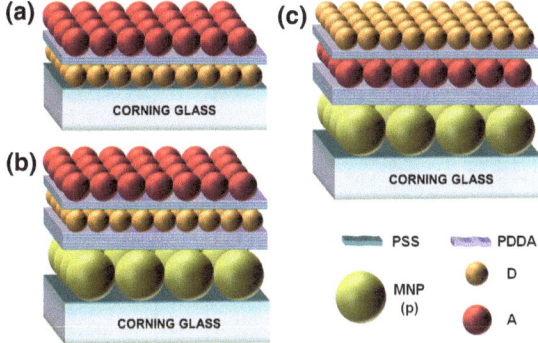

Fig. 13 Layered architectures of **a** conventional FRET, **b** plasmon-mediated FRET (PM-FRET) with plasmon coupling only to the donor quantum dots (where the plasmonic interaction with the acceptors is intentionally prevented), and **c** complimentary PM-FRET with plasmon coupling only to the acceptor quantum dots (while deliberately avoiding the plasmonic interaction with the donors). Reprinted (adapted) with permission from Ref. [184] (Copyright 2013 American Chemical Society)

acceptor-selective). Here, the authors demonstrated the ability to enable/disable the coupled plasmon-exciton (plexciton) formation distinctly at the donor site or at the acceptor site of their choice by using an optimized layer-by-layer assembled nanosystem composite of colloidal QD nanocrystal and metal nanoparticles (Fig. 13). The D-A exciton-exciton interaction involved in the FRET process was conserved while avoiding the plasmon-exciton coupling simultaneously to both sites of the donor and the acceptor pair.

In the case of donor-selective plexciton (Fig. 14), the donor QD lifetime decreases substantially from 1.33 to 0.29 ns as a result of the plasmon-coupling to the donors and the FRET-assisted exciton transfer from the donors to the acceptors, both of which shorten the donor lifetime. This causes an enhancement of the acceptor emission by a factor of 1.93 (Fig. 14b).

On the other hand, in the complimentary case of acceptor-selective plexciton (Fig. 15), the acceptor QDs emission is 2.70-fold enhanced as a result of the combined effects of the acceptor plasmon coupling and the FRET-assisted exciton feeding (Fig. 15b); this enhancement is larger than the acceptor emission enhancement of the donor-selective plexciton. In this work, the authors developed a theoretical model for the donor- and acceptor-selective plexcitons nonradiative energy transfer, which is in good agreement with the experimental results. In addition, the authors also demonstrated that it is possible to modify the donor or the acceptor of the FRET pairs selectively through plasmonics, without destroying the exciton-exciton interaction between them. Such modification of FRET mechanism with plasmonics holds great promise for FRET-driven nanophotonic device applications and FRET-based bioimaging. Selective control on the plexcitonic energy transfer will make it feasible to manipulate the detection signal and sensitivity of the desired donor or acceptor species selectively. Furthermore, in FRET

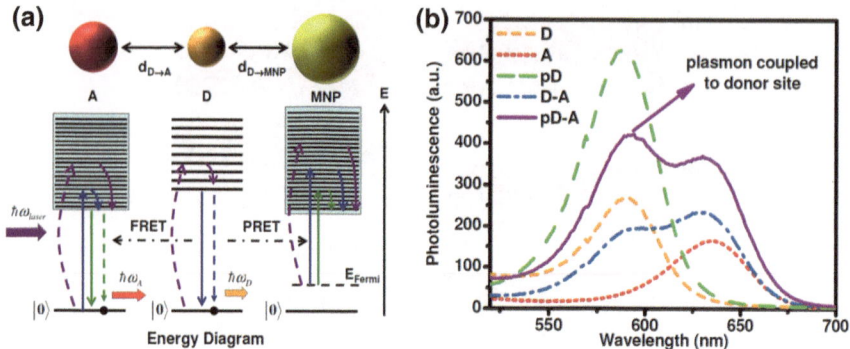

Fig. 14 **a** Schematic representation of a donor-acceptor (D-A) energy transfer pair in the case of plasmon coupling to only donor QD along with an energy band diagram in which the absorption process of the MNP/donor/acceptor, fast relaxation process, light emission process, energy transfer from the donor to the acceptor and the Coulomb interaction between the donor and acceptor pairs are shown. In the energy diagram, the discrete energy levels for the QDs are depicted, as well as the energy level for the localized plasmons within the continuous energy band of the MNP. **b** Photoluminescence (PL) spectra of the D (*dotted-orange*), A (*dotted-red*), D-A QD pair (*dashed-blue*) under Förster-type energy transfer, plasmon-coupled D (*dashed-green*) and FRET for the D-A QD pair when only the donor QD is coupled to MNP (*solid-magenta*). Reprinted (adapted) with permission from Ref. [184] (Copyright 2013 American Chemical Society)

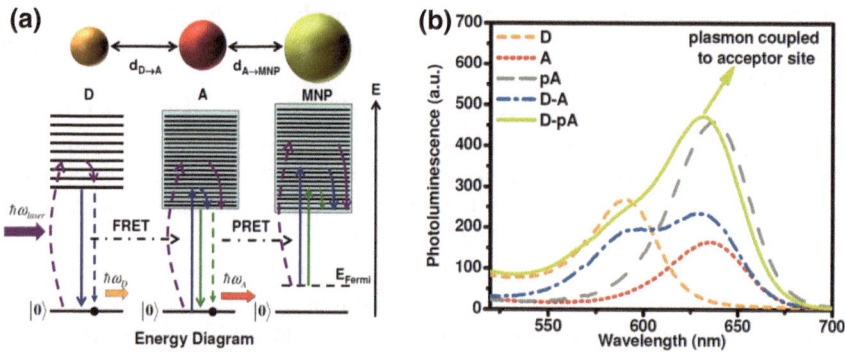

Fig. 15 **a** Schematic representation of a donor-acceptor (D-A) energy transfer pair in the case of plasmon coupling to only acceptor QD along with an energy band diagram with the absorption process of the MNP/donorQD/acceptorQD, fast relaxation process, light emission process, energy transfer from the donor to the acceptor and the Coulomb interaction between the donor and acceptor pairs are shown. In the energy diagram, the discrete energy levels for the QDs are depicted, as well as the energy level for the localized plasmons within the continuous energy band of the MNP. **b** Photoluminescence (PL) spectra of the D (*dotted-orange*), A (*dotted-red*), D-A QD pair (*dashed-blue*) under Förster-type energy transfer, plasmon-coupled A (*dashed-gray*) and FRET for the D-A QD pair when only the acceptor QD is coupled to MNP (*solid-yellow*). Reprinted (adapted) with permission from Ref. [184] (Copyright 2013 American Chemical Society)

studies where the energy transfer is used as the molecular ruler, plexcitonic energy transfer opens up the possibly to enhance the resolution of the measurement because of the enhanced energy transfer rate either using the donor or the acceptor of interest. [Reprinted (adapted) with permission from Ref. [184]. (Copyright 2013 American Chemical Society).]

5 Förster-type Nonradiative Energy Transfer to an Indirect Semiconductor (Silicon)

In this report, Yelik et al. [185] studied phonon-assisted Förster resonance energy transfer into indirect bandgap semiconductor (Si) using quantum dots emitters. The unusual temperature dependence of this energy transfer, which is measured using the QDs monolayers (donor) integrated on top of monocrystalline bulk silicon (acceptor), with each monolayer having a controlled separation distance (d) to the acceptor Si (Fig. 16), is predicted by a phonon-assisted exciton transfer model. The model includes the phonon-mediated optical properties of silicon, while considering the contribution from the multi-monolayer QDs into the energy transfer, which is derived to have a d^{-3} distance dependence. Figure 17 shows the energy diagram illustrating the exciton transfer from the donor QD to the acceptor silicon due to the Coulomb interaction between the donor-acceptor pair. The FRET efficiencies are experimentally observed to decrease at cryogenic temperatures, which were well explained by the model considering the phonon depopulation in the indirect bandgap acceptor, Si, together with the changes in the quantum yield of the donor.

In this study five different samples were used: 10-monolayer-equivalent QDs on top of either sapphire (reference), 0.0, 1.0, 2.0, or 4.0 nm Al_2O_3 separated SiO_2/Si substrates. Temperature-dependent amplitude-averaged fluorescence lifetimes of the QDs grafted on these five different substrates are presented in Fig. 18a–d for the cases of 0.0, 1.0, 2.0, and 4.0 nm thick Al_2O_3 separation, respectively. Green diamonds are the experimental reference lifetimes (QDs on sapphire—no FRET), which are corrected for the refractive index difference between sapphire and silicon. Black squares are the experimental lifetimes of the QDs when placed on a respective Al_2O_3 separation thickness. As a general trend, the lifetimes of the QDs are observed to increase with the decreasing temperature, which is in agreement with previous reports [186]. As depicted in Fig. 18 the models that include the phonon-assisted and full-temperature dependence for the energy transfer rates provide a more accurate fitting for the experimental data. Therefore, these findings indicate the assistance of phonons in FRET for the case of indirect bandgap acceptor such as silicon, the phonon assists for the case of optical absorption. These understandings is crucial for designing FRET-enabled sensitization of silicon for high-efficiency light-harvesting excitonic systems using nanoemitters. [Reprinted (adapted) with permission from Ref. [185]. (Copyright 2013 American Chemical Society).].

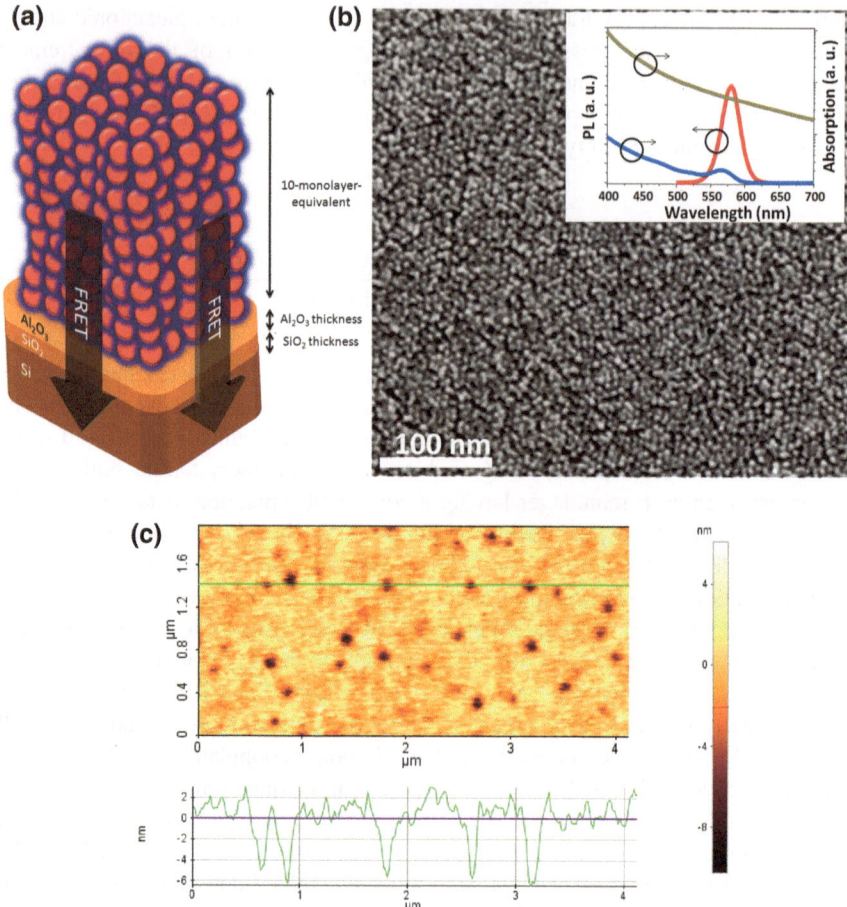

Fig. 16 **a** Schematic of the hybrid nanostructure of multimonolayer QDs and silicon separated by controlled Al$_2$O$_3$ separation *thickness*. Here, the native SiO$_2$ *thickness* is 1.65 nm. The *thickness* of Al$_2$O$_3$ film is varied from 0.0 to 4.0 nm. The QDs of the same monolayer are assumed to have the same exciton transfer contribution to bulk silicon. **b** SEM image of the QDs furnished on the Al$_2$O$_3$/SiO$_2$/Si structure. Inset shows optical absorption and PL spectra of the QDs (*black and green curves*, respectively) and absorption spectrum of silicon (*red curve*). **c** Atomic force microscopy image of the 10 monolayer-equivalent QD film on top of silicon with the height profile of the line shown inside the AFM image. Reprinted (adapted) with permission from Ref. [185]. (Copyright 2013 American Chemical Society)

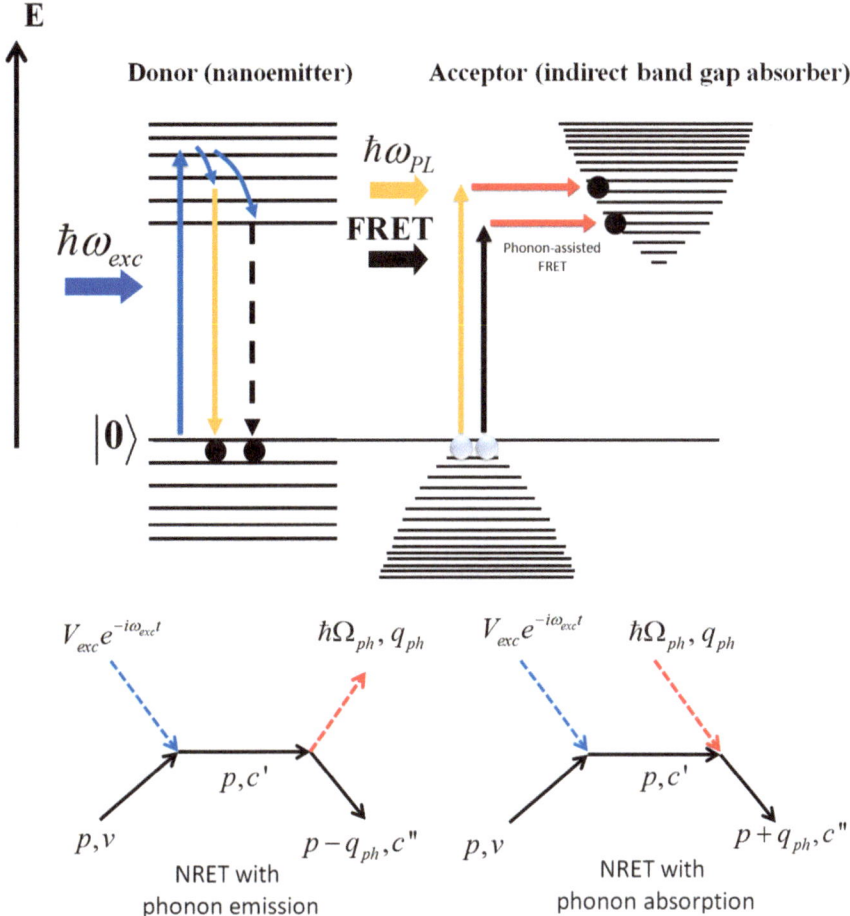

Fig. 17 **a** Energy diagram illustrating the exciton transfer from the donor QD to the acceptor silicon due to the Coulomb interaction between the donor-acceptor pair. The phonon-assisted process is shown as the lateral *arrows* to make up for the momentum mismatch in silicon. **b** Some of the Feynman diagrams for the phonon-assisted processes important for transfer of energy from a QD to an indirect-band semiconductor. These diagrams include phonon-emission processes in the conduction band. Reprinted (adapted) with permission from Ref. [185] (Copyright 2013 American Chemical Society)

6 Förster-type Nonradiative Energy Transfer and Coherent Transfer

In this work, Nizamoglu et al. [66] reported strong exciton migration with an efficiency of 83.3 % from a violet-emitting epitaxial quantum well to an energy gradient of layered green- and red-emitting quantum dots at room temperature (Fig. 19), enabled by the interplay between the exciton population and the

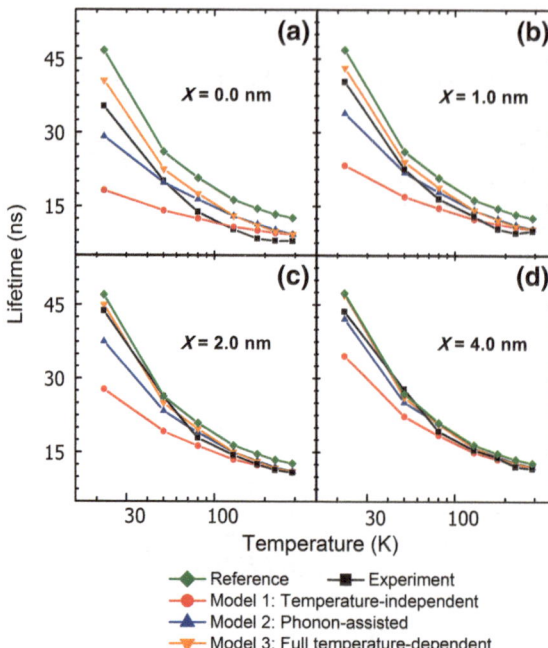

Fig. 18 Temperature dependence of fluorescence lifetime of the QDs integrated on top of **a** 0.0 nm, **b** 1.0 nm, **c** 2.0 nm, and **d** 4.0 nm thick Al_2O_3 layer on SiO_2/Si. *Black squares* are amplitude-averaged experiment lifetimes. *Green diamonds* are the lifetimes of the QDs on sapphire as the reference sample corrected for the refractive index difference with silicon. *Red circles* are the calculated lifetimes of the QDs using the "temperature-independent" energy transfer model. *Blue up-triangles* are the calculated lifetimes of the QDs using the energy transfer model which considers the temperature-dependent complex dielectric function of silicon, thus "phonon-assisted" model. *Orange down-triangles* are the calculated lifetimes of the QDs using the "phonon-assisted" energy transfer model with the additional inclusion of the temperature-dependent QY of the donor QDs; thus, it is called "full temperature-dependent" model. Reprinted (adapted) with permission from Ref. [185] (Copyright 2013 American Chemical Society)

depopulation of states in the optical near field. Based on the density matrix formalization of near-field interactions, the authors theoretically modelled and demonstrated that energy gradient significantly boosts the QW-QDs exciton transfer rate compared to using mono-dispersed QDs, which is in agreement with the observed experimental results. Figure 20a shows the numerical results for the exciton population. From this figure, it is observed that it takes about 1 ns for an exciton in the QW to be transferred to the second QD layer. Moreover, the nutation process between the first and second QD layer is observed; i.e., the exciton oscillate between the first excited state of both QDs because their energy levels are in resonance. The transfer efficiency from the QW to the last layer of QDs is about 45 %, as depicted in Fig. 20a (Inset). This result agrees with the experimental value of 50.7 %.

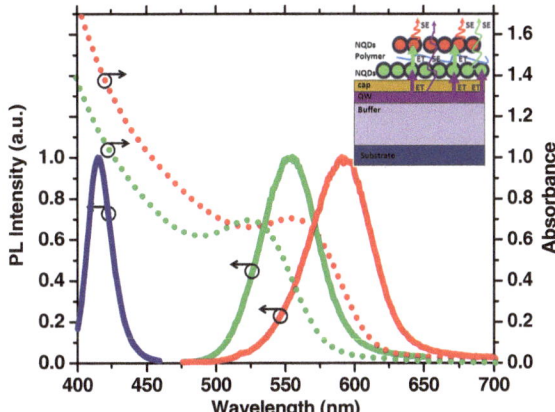

Fig. 19 Emission spectra of InGaN/GaN quantum well (*violet solid line*) and CdTe quantum dots (*green and red solid lines* for *green-* and *red-*emitting QDs, respectively) and absorption spectra of CdTe quantum dots (*green* and *red dotted lines* for *green* and *red-*emitting QDs, respectively). A schematic representing the hybrid structure of CdTe QDs bilayer integrated on the InGaN/GaN quantum well is shown. (ET and SE represent exciton transfer and spontaneous emission, respectively.) Reprinted with permission from Ref. [66] (Copyright 2012 AIP Publishing LLC)

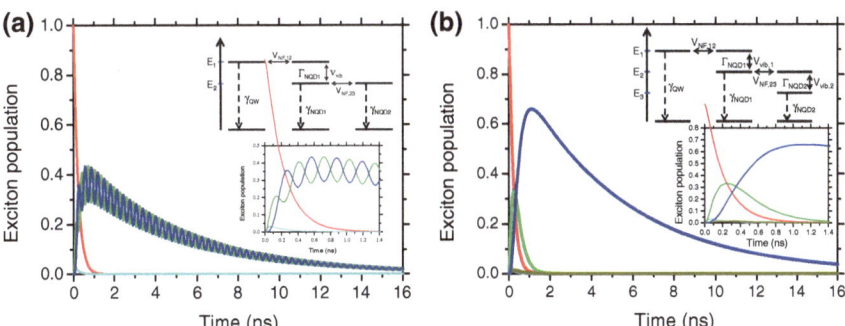

Fig. 20 **a** Calculated exciton population for the energy transfer from a QW to a bilayer of *red-*emitting QDs. *Red line* shows the exciton population of the first excited state in the QW. *Cyan line* depicts the exciton population of the second excited state for the first *red-*emitting QD layer; *green line* shows the exciton population of the first excited state of the first *red-*emitting QD layer; and the *blue line* depicts the exciton population of the first excited state of the second *red-*emitting QD layer. Inset: Energy diagram for the QW, the first QD layer and the second NQD layer. **b** Calculated exciton populations. *Red line* shows the exciton population of the first excited state in the QW. *Cyan line* depicts the exciton population of the second excited state for the *green* QD layer. *Green line* shows the exciton population of the first excited state of the *green* QD layer. *Dark yellow line* depicts the exciton population of the second excited state of the *red* QD layer. *Blue line* illustrates the exciton population of the first excited state of the *red* QD layer. Inset: Energy diagram for the QW, the first QD layer and the second QD layer. Reprinted with permission from Ref. [66] (Copyright 2012 AIP Publishing LLC)

Figure 20b shows the numerical results for the case that a green- and red-emitting QDs are placed on top of a QW. From this graph, it is observed that it takes between 1 and 2 ns for an exciton in the QW to be transfer to the red-emitting QD layer, similar to the previous case. Also, this plot illustrates the exciton population of the intermediate states. The transfer efficiency from the QW to the second layer is about 70 %, as depicted in Fig. 20b (inset). This result also agrees with the experimental value of 83.3 %. The advantage of energy gradient QDs system is that the exciton trapped in the trap states have the chance to be transferred to the adjacent energy level through near-field interaction. Therefore, the probability of exciton transfer to the first excited state of the second QD could significantly be improved. In summary, the results suggested that in order to have a stronger exciton transfer from the QW to the last QD layer, an energy difference between QD layers is needed. Similar results have been also reported in Refs. [187, 188]. [Reprinted with permission from Ref. [66]. (Copyright 2012 AIP Publishing LLC.)].

7   Conclusions

In summary, we discussed the excitonics of quantum confined dots and wires from their optical properties point of view, which is important for lighting and displays. First, we presented the excitonic interactions in the mixed hybrid systems of the QDs. Subsequently, current understanding of the excitonics in Qwires and some promising applications including Qwire lasers and LEDs were revised. After that, exciton-plasmon interactions and possible consequences on the enhanced optical properties and energy transfer abilities were also outlined. We finalize this chapter by highlighting several future perspectives and technological and scientific challenges towards excitonically engineered QD and Qwire systems for lighting and light-harvesting applications.

8   Future Challenges and Research Opportunities

Our understanding of the excitonic interactions in quantum confined structures has evolved considerably in recent years. Engineering the excitonic processes in QDs and QWs structures can boost the performance of existing devices such as colour converting QDs based LEDs or lead to novel devices such as Qwires lasers, which are all of high importance for efficient light generation systems. Nevertheless, the development of these excitonically controlled devices and systems will require a deeper understanding of the nanoscale interactions of these confined structures along with further developments in materials and processing techniques.

Excitonic energy transfer based enhancement of light generation has arisen as an important approach. QD LEDs, which employ QDs in electrically driven active layers, are promising for light generation in general. However, these QD-LEDs suffer from poor charge injection and subsequent low EQEs due to the organic insulating ligands. Excitonic injection can be used to address these current problems of charge-driven QD LEDs via pumping the QDs by FRET in addition to or instead of direct charge pumping. But, the challenge is to find a suitable material system that possesses good charge transport, exciton formation, and exciton transfer properties together. For example, quantum wells have been suggested as the appropriate exciton donors, since they have good charge transport and exciton formation properties, but it is difficult to integrate QDs close enough to quantum wells. Thus exciton transfer is limited in these hybrid systems [19, 63]. Another candidate as the proper donor material are organic materials including conjugated polymers and phosphorescent molecules, which are frequently used in organic LEDs efficiently. Nonetheless, a fully excitonic QD-LED has not been shown yet. Finally, Qwire-QD excitonic couplings for lighting might have promising prospects as compared to QW-QD systems. However, it is also a challenge to design a hybrid system of Qwires integrated with QDs for light-emitting purposes, which might find use in excitonically tunable LEDs and lasers. Excitonic emission in Qwire is suitable for achieving stimulated emission because the excitons have a strong tendency to radiatively recombine. However, ensuring stable excitons at room temperature is a challenge as well for various Qwires systems due to their small exciton binding energies, which requires high quantum confinement. Therefore, it is still technologically difficult and a challenge to synthesise Qwires with very small radii. As a result of these reasons, although it is very exciting to use FRET in active devices and systems as a driving mechanism, further developments are needed to make this approach practical.

References

1. B. Guzelturk, P.L. Hernandez-Martinez, Q. Zhang, Q. Xiong, H. Sun, X.W. Sun, A.O. Govorov, H.V. Demir, Excitonics of semiconductor quantum dots and wires for lighting and displays. Laser Photonics. Rev. **8**(1), 73–93 (2014)
2. Q. Sun, Y.A. Wang, L.S. Li, D. Wang, T. Zhu, J. Xu, C. Yang, Y. Li, Bright, multicolored light-emitting diodes based on quantum dots. Nat. Photonics **1**, 717–722 (2007)
3. F. Qian, S. Gradecak, Y. Li, C.-Y. Wen, C.M. Lieber, Core/Multishell nanowire heterostructures as multicolor, high-efficiency light-emitting diodes. Nano Lett. **5**, 2287–2291 (2005)
4. C.B. Murray, D.J. Norris, M.G. Bawendi, Synthesis and characterization of nearly monodisperse CdE (E = sulfur, selenium, tellurium) semiconductor nanocrystallites. J. Am. Chem. Soc. **115**, 8706–8715 (1992)
5. S.V. Gaponenko, *Introduction to Nanophotonics* (Cambridge University Press, 2010)
6. U. Woggon, S.V. Gaponenko, Excitons in quantum dots. Phys. Stat. Sol. (b) **189**, 285 (1995)
7. A.M. Morales, C.M. Lieber, A laser ablation method for the synthesis of semiconductor nanowires. Science **279**, 208–211 (1998)
8. J.T. Hu, T.W. Odom, C.M. Lieber, Chemistry and physics in one dimension: synthesis and properties of nanowires and nanotubes. Acc. Chem. Res. **32**, 435–455 (1999)
9. J.T. Hu, L.S. Li, W.D. Yang, L. Manna, L.W. Wang, A.P. Alivisatos, Linearly polarized emission from colloidal semiconductor nanorods. Science **292**, 2060–2063 (2001)

10. V.L. Colvin, M.C. Schlamp, A.P. Alivisatos, Light-emitting diodes made from cadmium selenide nanocrystals and a semiconducting polymer. Nature **370**, 354–357 (1994)
11. S. Coe, W.-K. Woo, M. Bawendi, V. Bulovic, Electroluminescence from single monolayers of nanocrystals in molecular organic devices. Nature **420**, 800–803 (2002)
12. W.U. Huynh, J.J. Dittmer, A.P. Alivisatos, Hybrid nanorod-polymer solar cells. Science **295**, 2425–2427 (2002)
13. H.V. Demir, S. Nizamoglu, T. Erdem, E. Mutlugun, N. Gaponik, A. Eychmüller, Quantum dot integrated LEDs using photonic and excitonic color conversion. Nano Today **6**, 632–647 (2011)
14. X. Duan, Y. Huang, Y. Cui, J. Wang, C.M. Lieber, Indium phosphide nanowires as building blocks for nanoscale electronic and optoelectronic devices. Nature **409**, 66–69 (2001)
15. M.H. Huang, S. Mao, H. Feick, H. Yan, Y. Wu, H. Kid, E. Weber, R. Russo, P. Yang, Room-temperature ultraviolet nanowire lasers. Science **292**, 1897–1899 (2001)
16. M.S. Gudiksen, L.J. Lauhon, J. Wang, D.C. Smith, C.M. Lieber, Growth of nanowire superlattice structrures for nanoscale photonics and electronics. Nature **415**, 617–620 (2002)
17. Y. Huang, X. Duan, C.M. Lieber, Nanowires for integrated multicolor nanophotonics. Small **1**, 142–147 (2005)
18. Y. Li, F. Qian, J. Xiang, C.M. Lieber, Nanowire electronic and optoelectronic devices. Mater. Today **9**, 18–27 (2006)
19. M. Achermann, M.A. Petruska, S. Kos, D.L. Smith, D.D. Koleske, V.I. Klimov, Energy transfer pumping of semiconductor nanocrystals using an epitaxial quantum well. Nature **429**, 642–646 (2004)
20. L. Novotny, B. Hecht, *Principles of Nano-optics* (Cambridge, 2006)
21. Z. Lin, H. Li, A. Franceschetti, M.T. Lusk, Efficient exciton transport between strongly quantum-confined silicon quantum dots. ACS Nano **6**, 4029–4038 (2012)
22. S.A. Crooker, J.A. Hollingsworth, S. Tretiak, V.I. Klimov, Spectrally resolved dynamics of energy transfer in quantum-dot assemblies: towards engineered energy flows in artificial materials. Phys. Rev. Lett. **89**, 186802 (2002)
23. S.W. Clark, J.M. Harbold, F.W. Wise, Resonant energy transfer in PbSe quantum dots. J. Phys. Chem. C **111**, 7302–7305 (2007)
24. S.F. Wuister, R. Koole, C.M. de Donega, A. Meijerink, Temperature-dependent energy transfer in cadmium telluride quantum dot solids. J. Phys. Chem. B. **109**, 5504–5508 (2005)
25. S. Nizamoglu, H.V. Demir, Excitation resolved color conversion efficiency of CdSe/ZnS core/shell quantum dot solids for hybrid white light emitting diodes. J. Appl. Phys. Lett. **105**, 083112 (2009)
26. B.N. Pal, Y. Ghosh, S. Brovelli, R. Laocharoensuk, V.I. Klimov, J.A. Hollingsworth, H. Htoon, Giant CdSe/CdS core/shell nanocrystal quantum dots as efficienct electroluminescent materials: strong influence of shell thickness on light-emitting diode performance. Nano Lett. **12**, 331–336 (2012)
27. A.L. Rogach, T.A. Klar, J.M. Lupton, A. Meijerink, J. Feldmann, Energy transfer with semiconductor nanocrystals. J. Mater. Chem. **19**, 1208–1221 (2009)
28. T. Franzl, T.A. Klar, S. Schietinger, A.L. Rogach, J. Feldmann, Exciton recycling in graded gap nanocrystal structures. Nano Lett. **4**, 1599–1603 (2004)
29. T.A. Klar, T. Franzl, A.L. Rogach, J. Feldmann, Super-efficient exciton funneling in layer-by-layer semiconductor nanocrystal structures. Adv. Mater. **17**, 769–773 (2005)
30. M. Achermann, M.A. Petruska, S.A. Crooker, V.I. Klimov, Picosecond energy transfer in quantum dot langmuir-Blodgett nanoassemblies. J. Phys. Chem. B **107**, 13782–13787 (2003)
31. S. Nizamoglu, H.V. Demir, Resonant nonradiative energy transfer in CdSe/ZnS core/shell nanocrystal solids enhances hybrid white light emitting diodes. Opt. Express **16**, 13961–13968 (2008)
32. T. Franzl, A. Shavel, A.L. Rogach, N. Gaponik, T.A. Klar, A. Eychmüller, J. Feldmann, High-rate unidirectional energy transfer in directly assembled CdTe nanocrystal bilayers. Small **1**, 392–395 (2005)

33. S. Nizamoglu, H.V. Demir, Förster resonance energy transfer enhanced color-conversion using colloidal semiconductor quantum dots for solid state lighting. Appl. Phys. Lett. **95**, 151111 (2009)
34. S. Nizamoglu, O. Akin, H.V. Demir, Quantum efficiency enhancement in nanocrystals using nonradiative energy transfer with optimized donor-acceptor ratio for hybrid LEDs. Appl. Phys. Lett. **94**, 243107 (2009)
35. K.E. Knowles, M.T. Frederick, D.B. Tice, A.J. Moris-Cohen, E.A. Weiss, Colloidal quantum dots: thinking outside the (particle-in-a) box. J. Phys. Chem. Lett. **3**, 18–26 (2012)
36. G.D. Scholes, D.L. Andrews, Resonance energy transfer and quantum dots. Phys. Rev. B. **72**, 125331 (2005)
37. A.O. Govorov, Spin-Förster transfer optically excited quantum dots. Phys. Rev. B. **71**, 155323 (2005)
38. S.Y. Kruchinin, A.V. Fedorov, A.V. Baranov, T.S. Perova, K. Berwick, Resonant energy transfer in quantum dots: frequency-domain fluorescent spectroscopy. Phys. Rev. B. **78**, 125311 (2008)
39. A. Nazir, Correlation-dependent coherent to incoherent transitions in resonant energy transfer dynamics. Phys. Rev. Lett. **103**, 146404 (2009)
40. C. Curutchet, A. Franceschetti, A. Zunger, G.D. Scholes, Examining Förster energy transfer for semiconductor nanocrystalline quantum dot donors and acceptors. J. Phys. Chem. C **112**, 13336–13341 (2008)
41. S. Halvini, A. Sitt, I. Hadar, U. Banin, Effect of nanoparticle dimensionality of fluorescence resonance energy transfer in nanoparticle-dye conjugated systems. ACS Nano **6**, 2758–2765 (2012)
42. E. Mutlugun, P.L. Hernandez-Martinez, C. Eroglu, Y. Coskun, T. Erdem, V.K. Sharma, E. Unal, S.K. Panda, S.G. Hickey, N. Gaponik, A. Eychmüller, H.V. Demir, Large-area (over 50 cm × 50 cm) freestanding films of colloidal InP/ZnS quantum dots. Nano Lett. **12**, 3986–3993 (2012)
43. D.A.B. Miller, Optoelectronic applications of quantum wells. Opt. Photonics News **1**, 7–15 (1990)
44. M.J. Bowers II, J.R. McBride, S.J. Rosenthal, White-light emission from magic-sized cadmium selenide nanocrystals. J. Am. Chem. Soc. **127**, 15378–15379 (2005)
45. H.-S. Chen, D.-M. Yeh, C.-F. Lu, C.-F. Huang, W.-Y. Shiao, J.-J. Huang, C.C. Yang, I.-S. Liu, W.-F. Su, White light generation with CdSe/ZnS nanocrystals coated on InGaN/GaN quantum-well blue/green two-wavelength light-emitting diode. IEEE Photonics. Technol. Lett. **18**, 1430–1432 (2006)
46. S. Nizamoglu, T. Ozel, E. Sari, H.V. Demir, White light generation using CdSe/ZnS core/shell nanocrystals hybridized with InGaN/GaN light emitting diodes. Nanotechnology **18**, 065709 (2007)
47. S. Nizamoglu, G. Zengin, H.V. Demir, Color-converting combinations of nanocrystal emitters for warm-white light generation with high color rendering index. Appl. Phys. Lett. **92**, 031102 (2008)
48. H.V. Demir, S. Nizamoglu, E. Mutlugun, T. Ozel, S. Sampra, N. Gaponik, A. Eychmuller, Tuning shades of white light with multi-color quantum-dot quantum-well emitters based on onion-like CdSe/ZnS heteronanocrystals. Nanotechnology **19**, 335203 (2008)
49. S. Nizamoglu, E. Mutlugun, T. Ozel, H.V. Demir, S. Sapra, N. Gaponik, A. Eychmüller, Dual-color emitting quantum-dot quantum-well CdSe/ZnS heteronanocrystals hybridized on InGaN/GaN light emitting diodes for high-quality white light generation. Appl. Phys. Lett. **92**, 113110 (2008)
50. C. Dang, J. Lee, Y. Zhang, J. Han, C. Breen, J.S. Steckel, S. Coe-Sullivan, A. Nurmikko, A wafer-level integrated integrated white light-emitting diode incorporating colloidal quantum dots as a nanocomposite luminescent materials. Adv. Mater. **24**, 5915–5918 (2012)
51. D. Basko, G.C. La Rocca, F.A. Bassani, V.M. Agranovich, Förster energy transfer from a semiconductor quantum well to an organic material overlayer. Eur. Phys. J. B **8**, 353–362 (1999)

52. S. Kos, M. Achermann, V.I. Klimov, D.L. Smith, Different regimes of förster-type energy transfer between an epitaxial quantum well and a proximal monolayer of semiconductor nanocrystals. Phys. Rev. B. **71**, 205309 (2005)
53. S. Blumstengel, S. Sadofev, C. Xu, J. Puls, F. Henneberger, Converting wannier into frenkel excitons in an inorganic/organic hybrid semiconductor structure. Phys. Rev. Lett. **97**, 237401 (2006)
54. G. Itskos, G. Heliotis, P.G. Lagoudakis, J. Lupton, N.P. Barradas, E. Alves, S. Pereira, I.M. Watson, M.D. Dawson, J. Feldmann, R. Murray, D.D.C. Bradley, Efficient dipole-dipole coupling of mott-wannier and frenkel excitons in (Ga, In)N quantum well/polyfluorene semiconductor structures. Phys. Rev. B. **76**, 035344 (2007)
55. C.R. Belton, G. Itskos, G. Heliotis, P.N. Stavrinou, P.G. Lagoudakis, J. Lupton, S. Pereira, E. Gu, C. Griffin, B. Guilhabert, I.M. Watson, A.R. Mackintosh, R.A. Pethrick, J. Feldmann, R. Murray, M.D. Dawson, D.D.C. Bradley, New light from hybrid organic-inorganic emitters. J. Phys. D Appl. Phys. **41**, 094006 (2008)
56. S. Chanyawadee, P.G. Lagoudakis, R.T. Harley, D.G. Lidzey, M. Henini, Nonradiative exciton energy transfer in hybrid organic-inorganic heterostructures. Phys. Rev. B **77**, 193402 (2008)
57. Y. Gladush, C. Piermarocchi, V. Agranovich, Dynamics of excitons and free carriers in hybrid organic-inorganic quantum well structures. Phys. Rev. B **84**, 205312 (2011)
58. J.J. Rindermann, G. Pozina, B. Monemar, L. Hultman, H. Amano, P.G. Lagoudakis, Dependence of resonance energy transfer on exciton dimensionality. Phys. Rev. Lett. **107**, 236805 (2011)
59. S. Lu, A. Madhukar, Nonradiative resonant excitation transfer from nanocrystal quantum dots to adjacent quantum channels. Nano Lett. **7**, 3443–3451 (2007)
60. S. Nizamoglu, E. Sari, J.-H. Baek, I.-H. Lee, H.V. Demir, Nonradiative resonance energy transfer directed from colloidal CdSe/ZnS quantum dots to epitaxial InGaN/GaN quantum wells for solar cells. Phys. Stat. Solid. RRL **4**, 178–180 (2010)
61. S. Chanyawadee, R.T. Harley, M. Henini, D.V. Talapin, P.G. Lagoudakis, Photocurrent enhancement in hybrid nanocrystal quantum-dot p-i-n photovoltaics devices. Phys. Rev. Lett. **102**, 077402 (2009)
62. S. Chanyawadee, P.G. Lagoudakis, R.T. Harley, M.D.B. Charlton, D.V. Talapin, H.W. Huang, C.-H. Lin, Increased color-conversion efficiency in hybrid light-emitting diodes utilizing non-radiative energy transfer. Adv. Mater. **22**, 602–606 (2010)
63. S. Nizamoglu, B. Guzelturk, D.-W. Jeon, I.-H. Lee, H.V. Demir, Efficient nonradiative energy transfer from InGaN/GaN nanopillars to CdSe/ZnS core/shell nanocrystals. Appl. Phys. Lett. **98**, 163108 (2011)
64. F. Zhang, J. Liu, G. You, C. Zhang, S.E. Mohney, M.J. Park, J.S. Kwak, Y. Wang, D.D. Koleske, J. Xu, Nonradiative energy transfer between colloidal quantum dot-phosphors and nanopillar nitride LEDs. Opt. Express **20**, 333–339 (2012)
65. B. Guzelturk, S. Nizamoglu, D.-W. Jeon, I.-H. Lee, H.V. Demir, *Strong nonradiative energy transfer from the nanopillars of quantum wells to quantum dots: effcient excitonic color conversion for light emitting diodes*, CLEO: Science and Innovations, CW1L (2012)
66. S. Nizamoglu, P.L. Hernandez-Martinez, E. Mutlugun, D.U. Karatay, H.V. Demir, Excitonic enhancement of nonradiative energy transfer from a quantum well in the optical near field of energy gradient quantum dots. Appl. Phys. Lett. **100**, 241109 (2012)
67. J. Lee, A.O. Govorov, N.A. Kotov, Bioconjugated superstructures of CdTe nanowires and nanoparticles: multistep cascade förster resonance energy transfer and energy channelling. Nano Lett. **5**, 2063–2069 (2005)
68. S. Lu, Z. Lingley, T. Asano, D. Harris, T. Barwicz, S. Guha, A. Madhukar, Photocurrent induced by nonradiative energy transfer from nanocrystal quantum dots to adjacent silicon nanowire conducting channels: Toward a new solar cell paradigm. Nano Lett. **9**, 4548–4552 (2009)
69. A. Dorn, D.B. Strasfeld, D.K. Harris, H.-S. Han, M.G. Bawendi, Using nanowires to extract excitons from a nanocrystal solid. ACS Nano **5**, 9028–9033 (2011)

70. P.L. Hernandez-Martinez, A.O. Govorov, Exciton energy transfer between nanoparticles and nanowires. Phys. Rev. B **78**, 035314 (2008)
71. J.Y. Kim, F.E. Osterloh, ZnO-CdSe nanoparticle clusters as directional photoemitters with tunable wavelength. J. Am. Chem. Soc. **127**, 10152–10153 (2005)
72. L.-J. Tzeng, C.-L. Cheng, Y.-F. Chen, Enhancement of band-edge emission induced by defect transition in the composite of ZnO nanorods and CdSe/ZnS quantum dots. Opt. Express **33**, 569–571 (2008)
73. J.-Y. Chang, T.G. Kim, Y.-M. Sung, Synergistic effects of SPR and FRET on the photoluminescence of ZnO nanorod heterostructures. Nanotechnology **22**, 425708 (2011)
74. J.M. Haremza, M.A. Hahn, T.D. Krauss, Attachment of single CdSe nanocrystals to individual single-walled carbon nanotubes. Nano Lett. **2**, 1253–1258 (2002)
75. X. Peng, J. Chen, J.A. Misewich, S.S. Wong, Carbon nanotube-nanocrystal heterostructures. Chem. Soc. Rev. **38**, 1076–1098 (2009)
76. M. Grzelczak, M.A. Correa-Duarte, V. Salgueirino-Maceira, M. Giersig, R. Diaz, L.M. Liz-Marzan, Photoluminescence quenching control in quantum dot-carbon nanotube composite colloids using a silica-shell spacer. Adv. Mater. **18**, 415–420 (2006)
77. V. Biju, T. Itoh, Y. Baba, M. Ishikawa, Quenching of photoluminescence in conjugates of quantum dots and single-walled carbon nanotube. J. Phys. Chem. B **110**, 26068–26074 (2006)
78. E. Shafran, B.D. Mangum, J.M. Gerton, Energy transfer from an individual quantum dot to a carbon nanotube. Nano Lett. **10**, 4049–4054 (2010)
79. B. Zebli, H.A. Vieyra, I. Carmeli, A. Hartschuh, J.P. Kotthaus, A.W. Holleitner, Optoelectronic sensitization of carbon nanotubes by CdTe nanocrystals. Phys. Rev. B **79**, 205402 (2009)
80. E. Morales-Navarez, B. Perez-Lopez, L.B. Pires, A. Merkoçi, Simple Förster resonance energy transfer evidence for the ultrahigh quantum dot quenching efficiency by graphene oxide compared to other carbon structures. Carbon **50**, 2987–2993 (2012)
81. S. Jander, A. Kornowski, H. Weller, Energy transfer from CdSe/CdS nanorods to amorphous carbon. Nano Lett. **11**, 5179–5183 (2011)
82. I.L. Medintz, H.T. Uyeda, E.R. Goldman, H. Mattoussi, Quantum dot bioconjugates for imaging labeling and sensing. Nat. Mater. **4**, 435–446 (2004)
83. C. Dang, J. Lee, C. Breen, J.S. Steckel, S. Coe-Sullivan, A. Nurmikko, Red, green and blue lasing enabled by single-exciton gain in colloidal quantum dot films. Nat. Nanotechnol. **7**, 335–339 (2012)
84. S.K. Saikin, A. Eisfeld, S. Valleau, A. Aspuru-Guzik, Photonics meets excitonics: natural and artificial molecular aggregates. Nanophotonics **2**, 21–38 (2013)
85. N. Tessler, V. Medvedev, M. Kazes, S.H. Kan, U. Banin, Efficient near-infrared polymer nanocrystal light-emitting diodes. Science **295**, 1056–1058 (2002)
86. M. Anni, L. Manna, R. Cingolani, D. Valerini, A. Creti, M. Lomascolo, Förster energy transfer from blue-emitting polymers to colloidal CdSe/ZnS core shell quantum dots. Appl. Phys. Lett. **85**, 4169–4171 (2004)
87. T.-W.F. Chang, S. Musikhin, L. Bakueva, L. Levina, M.A. Hines, P.W. Cyr, E.H. Sargent, Efficient excitation transfer from polymer to nanocrystals. Appl. Phys. Lett. **84**, 4295–4297 (2004)
88. J.H. Warner, A.R. Watt, E. Thomsen, N. Heckenberg, P. Meredith, H. Rubinsztein-Dunlop, Energy transfer dynamics of nanocrystal-polymer composites. J. Phys. Chem. B **109**, 9001–9005 (2005)
89. S.-K. Hong, Energy transfer by resonant dipole-dipole interaction from a conjugated polymer to a quantum-dot. Phys. E **28**, 66–75 (2005)
90. M.Y. Odoi, N.I. Hammer, K. Sill, T. Emrick, M.D. Barnes, Observation of enhanced energy transfer in individual quantum dot-oligophenylene vinylene nanostructures. J. Am. Chem. Soc. **128**, 3506–3507 (2006)

91. S. Kaufmann, T. Stöferle, N. Moll, R.F. Mahrt, U. Scherf, A. Tsami, D.V. Talapin, C.B. Murray, Resonant energy transfer within a colloidal nanocrystal polymer host system. Appl. Phys. Lett. **90**, 071108 (2007)
92. P.T.K. Chin, R.A.M. Hikmet, R.J. Janssen, Energy transfer in hybrid quantum dot light-emitting diodes. J. Appl. Phys. Lett. **104**, 013108 (2008)
93. T. Stöferle, U. Scherf, R.F. Mahrt, Energy transfer in hybrid organic/inorganic nanocomposites. Nano Lett. **9**, 453–456 (2009)
94. A.A. Lutich, G. Jiang, A.S. Susha, A.L. Rogach, F.D. Stefani, J. Feldmann, Energy transfer versus charge separation in type-II hybrid organic-inorganic nanocomposites. Nano Lett. **9**, 2636–2640 (2009)
95. A.A. Lutich, A. Pöschl, G. Jiang, F.D. Stefani, A.S. Susha, A.L. Rogach, J. Feldmann, Efficient energy transfer in layered hybrid organic/inorganic nanocomposites: a dual function of semiconductor nanocrystals. Appl. Phys. Lett. **96**, 083109 (2010)
96. E. Holder, N. Tessler, A.L. Rogach, Hybrid nanocomposite materials with organic and inorganic components for opto-electronic devices. J. Mater. Chem. **18**, 1064–1078 (2007)
97. N. Tomczak, D. Janczewski, M. Han, G.J. Vancso, Designer polymer-quantum dot architectures. Prog. Polym. Sci. **34**, 393–430 (2009)
98. P. Reiss, E. Couderc, J.D. Girolamo, A. Pron, Conjugated polymers/semiconductor nanocrystals hybrid materials—preparation, electrical transport properties and applications. Nanoscale **3**, 446–489 (2011)
99. B. Guzelturk, P.L. Hernandez-Martinez, V.K. Sharma, Y. Coskun, V. Ibrahimova, D. Tuncel, A.O. Govorov, X.W. Sun, Q. Xiong, H.V. Demir, Study of exciton transfer in dense quantum dot nanocomposites. Nanoscale **6**, 11387–11394 (2014)
100. P.O. Anikeeva, J.E. Halpert, M.G. Bawendi, V. Bulovic, Electroluminescence from a mixed red-green-blue colloidal quantum dot monolayer. Nano Lett. **7**, 2196–2200 (2007)
101. P.O. Anikeeva, J.E. Halpert, M.G. Bawendi, V. Bulovic, Quantum dot light-emitting devices with electroluminescence tunable over the entire visible spectrum. Nano Lett. **7**, 2532–2536 (2009)
102. P.O. Anikeeva, C.F. Madigan, J.E. Halpert, M.G. Bawendi, V. Bulovic, Electronic and excitonic processes in light-emitting devices based on organic materials and colloidal quantum dots. Phys. Rev. B **78**, 085434 (2008)
103. P. Jing, X. Yuan, W. Ji, M. Ikezawa, Y.A. Wang, X. Liu, L. Zhang, J. Zhao, Y. Masumoto, Shell-dependent energy transfer from 1,3,4-tris(n-phenylbenzimidazol-2, yl) benzene to CdSe core/shell quantum dots. J. Phys. Chem. C **114**, 19256–19262 (2010)
104. P. Jing, X. Yuan, W. Ji, M. Ikezawa, X. Liu, L. Zhang, J. Zhao, Y. Masumoto, Efficient energy transfer from hole transporting materials to CdSe-core CdS/ZnCdS/ZnS-multishell quantum dots in type II aligned blend films. Appl. Phys. Lett. **99**, 093106 (2011)
105. P.O. Anikeeva, C.F. Madigan, S.A. Coe-Sullivan, J.S. Steckel, M.G. Bawendi, V. Bulovic, Photoluminescence of CdSe/ZnS core/shell quantum dots enhanced by energy transfer from a phosphorescent donor. Chem. Phys. Lett. **424**, 120–125 (2006)
106. G. Cheng, M. Mazzeo, A. Rizzo, Y. Li, Y. Duan, G. Gigli, White light-emitting devices based on the combined emission from red CdSe/ZnS quantum dots, green phosphorescent and blue fluorescent organic molecules. Appl. Phys. Lett. **94**, 243506 (2009)
107. Y.Q. Zhang, X.A. Cao, Electroluminescence of green CdSe/ZnS quantum dots enhanced by harvesting excitons from phosphorescent molecules. Appl. Phys. Lett. **97**, 235115 (2010)
108. G. Cheng, W. Lu, Y. Chen, C.-M. Che, Hybrid light-emitting devices based on phosphorescent platinum (II) complex sensitized CdSe/ZnS quantum dots. Opt. Express **37**, 1109–1111 (2012)
109. D.M. Willard, L.L. Carillo, J. Jung, A.V. Orden, CdSe-ZnS quantum dots as resonance energy transfer donors in a model protein-protein binding assay. Nano Lett. **1**, 469–474 (2001)
110. A.R. Clapp, I.L. Medintz, J.M. Mauro, B.R. Fisher, M.G. Bawendi, H. Mattoussi, Fluorescence resonance energy transfer between quantum dot donors and dye-labeled protein acceptors. J. Am. Chem. Soc. **126**, 301–310 (2004)

111. I.L. Medintz, A.R. Clapp, H. Mattoussi, E.R. Goldman, B. Fisher, J.M. Mauro, Self-assembled nanoscale biosensors based on quantum FRET donors. Nat. Mater. **2**, 630–638 (2003)

112. I.L. Medintz, J.H. Konnert, A.R. Clapp, I. Stanish, M.E. Twigg, H. Mattoussi, J.M. Mauro, J. R. Deschamps, A fluorescence resonance energy transfer-derived structure of a quantum dot-protein bioconjugate nanoassembly. Proc. Natl. Acad. Sci. **101**, 9612–9617 (2004)

113. I.L. Medintz, H. Mattoussi, Quantum dot-based resonance energy transfer and its growing applications in biology. Phys. Chem. Chem. Phys. **11**, 17–45 (2009)

114. W.R. Algar, D. Wegner, A.L. Huston, J.B. Blanco-Canosa, M.H. Stewart, A. Armstrong, P. E. Dawson, N. Hildebrandt, I.L. Medintz, Quantum dots as simultaneous acceptors and donors in time-gated Förster resonance energy transfer relays: characterization and biosensing. J. Am. Chem. Soc. **134**, 1876–1891 (2012)

115. M.-K. So, C. Xu, A.M. Loening, S.S. Gambhir, J. Rao, Self-illuminating quantum dot conjugates for in vivo imaging. Nat. Biotech. **24**, 339–343 (2006)

116. H. Yao, Y. Zhang, F. Xiao, Z. Xia, J. Rao, Quantum dot/bioluminescence resonance energy transfer based highly sensitive detection of proteases. Angew. Chem. Int. Ed. **46**, 4346–4349 (2007)

117. X. Huang, L. Li, H. Qian, C. Dong, J. Ren, A resonance energy transfer between chemiluminescent donors and luminescent quantum-dot acceptors (CRET). Angew. Chem. **118**, 5264–5267 (2006)

118. M.-S. Wu, H.-W. Shi, J.-J. Xu, H.-Y. Chen, CdS quantum dots/Ru(bpy)$_3^{2+}$ electrochemiluminescence resonance energy transfer system for sensitive cytosensing. Chem. Commun. **47**, 7752–7754 (2011)

119. L. Li, M. Li, Y. Sun, J. Li, L. Sun, G. Zou, X. Zhang, W. Jin, Electrochemiluminescence resonance energy transfer between an emitter electrochemically generated by luminol as the donor and luminescent quantum dots as the acceptor and its biological applications. Chem. Commun. **47**, 8292–9294 (2011)

120. M. Artemyev, E. Ustinovich, I. Nabiev, Efficiency of energy transfer from organic dye molecules to CdSe-ZnS nanocrystals: nanorods versus nanodots. J. Am. Chem. Soc. **131**, 8061–8065 (2009)

121. E. Mutlugun, S. Nizamoglu, H.V. Demir, Highly efficient nonradiative energy transfer using charged CdSe/ZnS nanocrystals for light-harvesting in solution. Appl. Phys. Lett. **95**, 033106 (2009)

122. S. Sarkar, R. Bose, S. Jana, N.R. Jana, N. Pradhan, Doped semiconductor nanocrystals and organic dyes: an efficient and greener FRET systems. J. Phys. Chem. Lett. **1**, 636–640 (2010)

123. M. Achermann, S. Jeong, L. Balet, G.A. Montano, J.A. Hollingsworth, Efficient quantum dot-quantum dot and quantum dot-dye energy transfer in biotemplated assemblies. ACS Nano **5**, 1761–1768 (2011)

124. S. Halivni, A. Sitt, I. Hadar, U. Banin, Effect of nanoparticles dimensionality on fluorescence resonance energy transfer in nanoparticle—dye conjugated systems. ACS Nano **6**, 2758–2765 (2012)

125. T. Takagahara, Effects of dielectric confinement and electron-hole exchange interaction on excitonic states in semiconductor quantum dots. Phys. Rev. B **47**, 4569–4584 (1993)

126. T. Someya, H. Akiyama, H. Sakaki, Enhanced binding energy of one-dimensional excitons in quantum wires. Phys. Rev. Lett. **76**, 2965–2968 (1996)

127. E.A. Muljarov, E.A. Zhukov, V.S. Dneprovskii, Y. Masumoto, Dielectrically enhanced excitons in semiconductor-insulator quantum wires: Theory and experiment. Phys. Rev. B **62**, 7420–7432 (2000)

128. Q. Li, X. Gong, C. Wang, J. Wang, K. Ip, S. Hark, Size-dependent periodically twinned ZnSe nanowires. Adv. Mater. **16**, 1436–1440 (2004)

129. R. Chen, D. Li, B. Liu, Z. Peng, G.G. Gurzadyan, Q.H. Xiong, H. Sun, Optical and excitonic properties of crystalline ZnS nanowires: towards efficient ultraviolet emission at room temperature. Nano Lett. **10**, 4956–4961 (2010)

130. M.H. Huang, S. Mao, H. Feick, H. Yan, Y. Wu, H. Kind, E. Weber, R. Russo, P. Yang, Room-temperature ultraviolet nanowire nanolaser. Science **292**, 1897–1899 (2001)

131. R. Agarwal, C.J. Barrelet, C.M. Lieber, Lasing in single cadmium sulfide nanowire optical cavities. Nano Lett. **5**, 917–920 (2005)

132. R. Chen, M.I.B. Utama, Z. Peng, B. Peng, Q.H. Xiong, H. Sun, Excitonic processes and near-infrared coherent random lasing in vertically aligned CdSe nanowires. Adv. Mater. **23**, 1404–1408 (2011)

133. X. Xu, Y. Zhao, E.J. Sie, Y. Lu, B. Liu, S.A. Ekahana, X. Ju, Q. Jiang, J. Wang, H. Sun, T.C. Sum, C.H.A. Huan, Y.P. Feng, Q.H. Xiong, Dynamics of bound exciton complexes in CdS nanobelts. ACS Nano **5**, 3660–3669 (2011)

134. L.J. Lauhon, M.S. Gudiksen, C.M. Lieber, Semiconductor nanowire heterostructures. Phil. Trans. R. Soc. Lond. A **362**, 1247–1260 (2004)

135. F. Qian, Y. Li, S. Gradecak, H.-G. Park, Y. Dong, Y. Ding, Z.L. Wang, C.M. Lieber, Multi-quantum-well nanowire heterostructures for wavelength-controlled lasers. Nat. Mater. **7**, 701–706 (2008)

136. F. Qian, M. Brewster, S.K. Lim, Y. Ling, C. Greene, O. Laboutin, J.W. Johnson, S. Gradecak, Y. Cao, Y. Li, Controlled synthesis of AlN/GaN multiple quantum well nanowire structures and their optical properties. Nano Lett. **12**, 3344–3350 (2012)

137. R. Könenkamp, R.C. Word, C. Schlegel, Verical nanowire light-emitting diode. Appl. Phys. Lett. **85**, 7004–7006 (2004)

138. R. Könenkamp, R.C. Word, M. Godinez, Ultraviolet electroluminescence from ZnO/polymer heterojunction light-emitting diodes. Nano Lett. **5**, 2005–2008 (2005)

139. J. Bao, M.A. Zimmler, F. Capasso, Broadband ZnO single-nanowire light-emitting diode. Nano Lett. **6**, 1719–1722 (2006)

140. A. Nadarajah, R.C. Word, J. Meiss, R. Könenkamp, Flexible inorganic nanowire light-emitting diodes. Nano Lett. **8**, 534–537 (2008)

141. X.-M. Zhang, M.-Y. Lu, Y. Zhang, L.-J. Chen, Z.L. Wang, Fabrication of a high-brightness blue-light-emitting diode using a ZnO-nanowire array grown on p-GaN thin film. Adv. Mater. **21**, 2767–2770 (2009)

142. O. Lupan, T. Pauporte, B. Viana, Low-voltage UV-electroluminescence from ZnO-nanowire array/p-GaN light-emitting diode. Adv. Mater. **22**, 3298–3302 (2010)

143. M.A. Zimmler, T. Voss, C. Ronning, F. Capasso, Exciton-related electroluminescence from ZnO nanowire light-emitting diodes. Appl. Phys. Lett. **94**, 241120 (2009)

144. B. Yan, R. Chen, W. Zhou, J. Zhang, H. Sun, H. Gong, T. Yu, Localized suppression of longitudinal-optical-phonon-exciton coupling in bent ZnO nanowires. Nanotechnology **21**, 445706 (2010)

145. Q. Yang, W. Wang, S. Xu, Z.L. Wang, Enhancing light emission of ZnO microwire-based diodes by piezo-phototronic effect. Nano Lett. **11**, 4012–4017 (2011)

146. S. Nanot, E.H. Haroz, J.-H. Kim, R.H. Hauge, J. Kono, Optoelectronic properties of single-walled carbon nanotubes. Adv. Mater. **24**, 4977–4994 (2012)

147. C.D. Spataru, S. Ismail-Beigi, L.X. Benedict, S.G. Louie, Excitonic effects and optical spectra of single-walled carbon nanotubes. Phys. Rev. Lett. **92**, 077402 (2004)

148. V. Perebeinos, J. Tersoff, P. Avouris, Scaling of excitons in carbon nanotubes. Phys. Rev. Lett. **92**, 257402 (2004)

149. F. Wang, G. Dukovic, L.E. Brus, T.F. Heinz, The optical resonances in carbon nanotubes arise from excitons. Science **308**, 838–843 (2005)

150. I.V. Bondarev, L.M. Woods, K. Tatur, Strong exciton-plasmon coupling in semiconducting carbon nanotubes. Phys. Rev. Lett. **80**, 085407 (2009)

151. D.J. Bindl, M.-Y. Wu, F.C. Prehn, M.S. Arnold, Efficiently harvesting excitons from electronic type-controlled semiconducting carbon nanotube films. Nano Lett. **11**, 455–460 (2010)

152. A.J. Siitonen, D.A. Tsyboulski, S.M. Bachilo, R.B. Weisman, Surfactant-dependent exciton mobility in single-walled carbon nanotubes studied by single-molecule reactions. Nano Lett. **10**, 1595–1599 (2010)

153. S. Wang, M. Kahfizov, X. Tu, M. Zheng, T.D. Krauss, Multiple exciton generation in single-walled carbon nanotubes. Nano Lett. **10**, 2381–2386 (2010)
154. A. Högele, C. Galland, M. Winger, A. Imamoglu, Photon antibunching in the photoluminescence spectra of a single carbon nanotube. Phys. Rev. Lett. **100**, 217401 (2008)
155. T. Mueller, M. Kinoshita, M. Steiner, V. Perebeinos, A.A. Bol, D.B. Farmer, P. Avouris, Efficient narrow-band light emission from a single carbon nanotube p-n diode. Nat. Nanotechnol. **5**, 27–31 (2009)
156. S. Wang, Q. Zheng, L. Yang, Z. Zhang, Z. Wang, T. Pei, L. Ding, X. Liang, M. Gao, Y. Li, L.-M. Peng, High-performance carbon nanotube light-emitting diodes with asymmetric contacts. Nano Lett. **11**, 23–29 (2011)
157. X. Xie, A.E. Islam, M.A. Wahab, L. Ye, X. Ho, M.A. Alam, J.A. Rogers, Electroluminescence in aligned arrays of single-walled carbon nanotubes with asymmetric contacts. ACS Nano **6**(9), 7981–7988 (2012)
158. P.H. Tan, A.G. Rozhin, T. Hasan, P. Hu, V. Scardaci, W.I. Milne, A.C. Ferrari, Photoluminescence spectroscopy of carbon nanotube bundles: evidence for exciton energy transfer. Phys. Rev. Lett. **99**, 137402 (2007)
159. H. Qian, C. Georgi, N. Anderson, A.A. Green, M.C. Hersam, L. Novotny, A. Hartschuh, Exciton energy transfer in pairs of single-walled carbon nanotubes. Nano Lett. **8**, 1363–1367 (2008)
160. H. Qian, C. Georgi, N. Anderson, A.A. Green, M.C. Hersam, L. Novotny, A. Hartschuh, Exciton transfer and propagation in carbon nanotubes studied by near-field optical microscopy. Phys. Stat. Sol. (b) **245**, 2243–2246 (2008)
161. J. Lefebvre, P. Finnie, Photoluminescence and förster resonance energy transfer in elemental bundles of single-walled carbon nanotubes. J. Phys. Chem. C **113**, 7536–7540 (2009)
162. H. Hirori, K. Matsuda, Y. Kanemitsu, Exciton energy transfer between the inner and outer tubes in double-walled carbon nanotubes. Phys. Rev. B **78**, 113409 (2008)
163. T. Koyama, Y. Asada, N. Hikosaka, Y. Miyata, H. Shinohara, A. Nakamura, Ultrafast exciton energy transfer between nanoscale coaxial cylinders: intertube transfer and luminescence quenching in double-walled carbon nanotubes. ACS Nano **5**, 5881–5887 (2011)
164. A.O. Govorov, G.W. Bryant, W. Zhang, T. Skeini, J. Lee, N.A. Kotov, J.M. Slocik, R.R. Naik, Exciton-plasmon interaction and hybrid excitons in semiconductor—metal nanoparticle assemblies. Nano Lett. **6**, 984–994 (2006)
165. N.T. Fofang, T.-H. Park, O. Neumann, N.A. Mirin, P. Nordlander, N.J. Halas, Plexcitonic particles: plasmon-exciton coupling in nanoshell – J-aggregate compexes. Nano Lett. **8**, 3481–3487 (2008)
166. P.D. Artuso, G.W. Bryant, Optical response of strongly coupled quantum dot—metal nanoparticle systems: double peaked fano structure and bistability. Nano Lett. **8**, 2106–2111 (2008)
167. D.E. Gomez, K.C. Vernon, P. Mulvaney, T.J. Davis, Surface plasmon mediated strong exciton—photon coupling in semiconductor nanocrystals. Nano Lett. **10**, 274–278 (2010)
168. A. Manjavacas, F.J.G. de Abajo, P. Nordlander, Quantum plexcitonics: strongly interaction plasmons and excitons. Nano Lett. **11**, 2318–2323 (2011)
169. O. Kulakovich, N. Strekal, A. Yaroshevich, S. Maskevich, S. Gaponenko, I. Nabiev, U. Woggon, M. Artemyev, Enhanced luminescence of CdSe quantum dots on gold colloids. Nano Lett. **2**, 1449–1452 (2002)
170. T.E. Jennings, J.C. Schlatterer, M.P. Singh, N.L. Greenbaum, G.F. Strouse, NSET molecular beacon analysis of hammerhead RNA substrate binding and catalysis. Nano Lett. **6**, 1218–1324 (2006)
171. O.G. Tovmachenko, C. Graf, D.J. van den Heuvel, A. van Blaaderen, H.C. Gerritsen, Fluorescence enhancement by metal-core/silica-shell nanoparticles. Adv. Mater. **18**, 91–95 (2006)

172. T. Pons, I.L. Medintz, K.E. Sapsford, S. Higashiya, A.F. Grimes, D.S. English, H. Matoussi, On the quenching of semiconductor quantum dot photoluminescence by proximal gold nanoparticles. Nano Lett. **7**, 3157–3164 (2007)
173. B. Peng, Q. Zhang, X. Liu, Y. Li, H.V. Demir, C.H.A. Huan, T.Z. Sum, Q.H. Xiong, Fluorophore-doped core-multishell spherical plasmonic nanocavities: resonant energy transfer toward a loss compensation. ACS Nano **6**, 6250–6259 (2012)
174. R. Schreiber, J. Do, E.-M. Roller, T. Zhang, V.J. Schüller, P.C. Nickels, J. Feldmann, T. Liedl, Hierarchical assembly of metal nanoparticles, quantum dots and organic dyes using DNA origami scaffolds. Nat. Nanotechnol. **9**, 74–78 (2014)
175. V. Biju, T. Itoh, A. Anas, A. Sujith, M. Ishikawa, Semiconductor quantum dots and metal nanoparticles: syntheses, optical properties, and biological applications. Anal. Bioanal. Chem. **391**, 2469–2495 (2008)
176. J. Lee, A.O. Govorov, J. Dulka, N.A. Kotov, Bioconjugates of CdTe nanowires and Au nanoparticles: plasmon-exciton interactions, luminescence enhancement and collective effects. Nano Lett. **4**, 2323 (2004)
177. J. Lee, T. Javed, T. Skeini, A.O. Govorov, G.W. Bryant, N.A. Kotov, Bioconjugated Ag nanoparticles and CdTe nanowires: luminescence enhancement in metamaterials due to field-enhanced light absorption. Angew. Chem. **45**, 4819 (2006)
178. A.O. Govorov, J. Lee, N.A. Kotov, Theory of plasmon-enhanced Förster resonance energy transfer in optically excited semiconductor and metal nanoparticles. Phys. Rev. B **76**, 125308 (2007)
179. V. Faessler, C. Hrelescu, A.A. Lutich, L. Osinkina, S. Mayilo, F. Jackel, J. Feldmann, Accelerating fluorescence resonance energy transfer with plasmonic nanoresonators. Chem. Phys. Lett. **508**, 67–70 (2011)
180. M. Lunz, V.A. Gerard, Y.K. Gunko, V. Lesnyak, N. Gaponik, A.S. Susha, A.L. Rogach, A. L. Bradley, Surface plasmon enhanced energy transfer between donor and acceptor CdTe nanocrystal quantum dot monolayers. Nano Lett. **11**, 3341–3345 (2011)
181. T. Ozel, S. Nizamoglu, M.A. Sefunc, O. Samarskaya, I.O. Ozel, E. Mutlugun, V. Lesnyak, N. Gaponik, A. Eychmuller, S.V. Gaponenko, H.V. Demir, Anisotropic emission from multilayered plasmon resonator nanocomposites of isotropic semiconductor quantum dots. ACS Nano **5**, 1328–1334 (2011)
182. P. Andrew, W.L. Barnes, Förster energy transfer in an optical microcavity. Science **290**, 785–788 (2000)
183. P. Andrew, W.L. Barnes, Energy transfer across a metal film mediated by surface plasmon polaritons. Science **306**, 1002–1005 (2004)
184. T. Ozel, P.L. Hernandez-Martinez, E. Mutlugun, O. Akin, S. Nizamoglu, I.O. Ozel, Q. Zhang, Q. Xiong, H.V. Demir, Observation of selective plasmon-exciton coupling in nonradiative energy transfer: donor-selective versus acceptor-selective plexcitons. Nano Lett. **13**(7), 3065–3072 (2013)
185. A. Yeltik, B. Guzelturk, P.L. Hernandez-Martinez, A.O. Govorov, H.V. Demir, Phonon-assisted exciton transfer into silicon using nanoemitters: the role of phonons and temperature effects in FRET. ACS Nano **7**(12), 10492–10501 (2013)
186. O. Labeau, P. Tamarat, B. Lounis, Temperature dependence of the luminescence lifetime of single CdSe-ZnS quantum dots. Phys. Rev. Lett. **90**, 257404 (2003)
187. T. Franzl, T.A. Klar, S. Schietinger, A.L. Rogach, J. Feldmann, Exciton recycling in graded gap nanocrystal structures. Nano Lett. **4**, 1599 (2004)
188. T.A. Klar, T. Franzl, A.L. Rogach, J. Feldmann, Super-efficient exciton funneling in layer-by-layer semiconductor nanocrystal structures. Adv. Mater. **17**, 769 (2005)